The Magic Numbers
of the Professor

© 2007 by The Mathematical Association of America, Inc.

ISBN 10: 0-88385-557-7

ISBN 13: 978-0-88385-557-7

Library of Congress Control Number: 2006933947

Current Printing (last digit):

10 9 8 7 6 5 4 3 2 1

The Magic Numbers of the Professor

Owen O'Shea

and

Underwood Dudley

Published and Distributed by
The Mathematical Association of America

SPECTRUM SERIES

The Spectrum Series of the Mathematical Association of America was so named to reflect its purpose: to publish a broad range of books including biographies, accessible expositions of old or new mathematical ideas, reprints and revisions of excellent out-of-print books, popular works, and other monographs of high interest that will appeal to a broad range of readers, including students and teachers of mathematics, mathematical amateurs, and researchers.

777 Mathematical Conversation Starters, by John de Pillis

99 Points of Intersection: Examples—Pictures—Proofs, by Hans Walser. Translated from the original German by Peter Hilton and Jean Pedersen

aha! A two volume collection: aha! Gotcha and aha! Insight, by Martin Gardner

All the Math That's Fit to Print, by Keith Devlin

Carl Friedrich Gauss: Titan of Science, by G. Waldo Dunnington, with additional material by Jeremy Gray and Fritz-Egbert Dohse

The Changing Space of Geometry, edited by Chris Pritchard

Circles: A Mathematical View, by Dan Pedoe

Complex Numbers and Geometry, by Liang-shin Hahn

Cryptology, by Albrecht Beutelspacher

The Early Mathematics of Leonhard Euler, by C. Edward Sandifer

The Edge of the Universe: Celebrating 10 Years of Math Horizons, edited by Deanna Haunsperger and Stephen Kennedy

Five Hundred Mathematical Challenges, Edward J. Barbeau, Murray S. Klamkin, and William O. J. Moser

The Genius of Euler: Reflections on his Life and Work, edited by William Dunham

The Golden Section, by Hans Walser. Translated from the original German by Peter Hilton, with the assistance of Jean Pedersen.

I Want to Be a Mathematician, by Paul R. Halmos

Journey into Geometries, by Marta Sved

JULIA: A Life in Mathematics, by Constance Reid

R. L. Moore: Mathematician and Teacher, by John Parker

The Lighter Side of Mathematics: Proceedings of the Eugène Strens Memorial Conference on Recreational Mathematics & Its History, edited by Richard K. Guy and Robert E. Woodrow

Lure of the Integers, by Joe Roberts

The Magic Numbers of the Professor, by Owen O'Shea and Underwood Dudley

Magic Tricks, Card Shuffling, and Dynamic Computer Memories: The Mathematics of the Perfect Shuffle, by S. Brent Morris

Martin Gardner's Mathematical Games: The Entire Collection of His Scientific American Columns

The Math Chat Book, by Frank Morgan

Mathematical Adventures for Students and Amateurs, edited by David Hayes and Tatiana Shubin. With the assistance of Gerald L. Alexanderson and Peter Ross

Mathematical Apocrypha, by Steven G. Krantz

Mathematical Apocrypha Redux, by Steven G. Krantz

Mathematical Carnival, by Martin Gardner

MAA Service Center
P.O. Box 91112
Washington, DC 20090-1112
800-331-1622   FAX: 301-206-9789

Foreword by Martin Gardner

Did you ever notice that remarkable coincidence? Bernard Shaw is 61 years old, H. G. Wells is 51, G. K. Chesterton is 41, you're 31 and I'm 21— all the great authors of the world in arithmetical progression.

—F. Scott Fitzgerald in a 1928 letter to Shane Leslie.

I never had the pleasure of meeting Owen O'Shea—he lives in Cobh (pronounced Cove!), Ireland, I in the United States. But over the years we had a stimulating correspondence. O'Shea's letters were crammed with the most astonishing mathematical curiosities. Most of them were his own discoveries, others were found in obscure books and periodicals. It soon became clear that O'Shea had a rare and uncanny ability to uncover such curios, not only in number theory but also in other branches of mathematics. Now he has happily gathered together a wealth of such oddities in a single volume, that I believe will become a classic reference on such things.

O'Shea's range is awesome. His garden of weird coincidences includes the numerology of major wars, and disasters such as Pearl Harbor and the 9/11 tragedies. There are miraculous coincidences involving numbers, equations, geometry, dice, cards, dart scoring, currency, magic squares, the calendar, geography, 666, pi, curiosities in James Joyce's *Ulysses* and the King James Bible, new correlations between the lives of Abraham Lincoln and John Kennedy, word play on the names of people in the news, anagrams, palindromes, and scores of other linguistic amusements.

When encountering unlikely coincidences there is a strong temptation to suspect that something other than chance is involved. Carl Jung popularized the term *synchronicity* to suggest that paranormal forces may somehow be at work. Writer Arthur Koestler, who saw the evils of Stalinism long before Communism's gullible fellow travelers saw them, took Jung's synchronicity with utmost seriousness in his book *The Roots of Coincidence*. Edgar Allan Poe, in his story *The Mystery of Marie Roget*, expressed the paranormal temptation this way:

> There are few persons, even among the calmest thinkers, who have not occasionally been startled into a vague yet thrilling half-credence in the supernatural, by coincidences of so seemingly marvellous a character that as, mere coincidences, the intellect has been unable to receive them.

An awareness of the enormously different ways that coincidences can occur goes all the way back to Aristotle. It is extremely probable, he wrote, that improbable events occur. Nevertheless, there is something spooky about the seemingly unlikely coincidences that O'Shea has packed into his entertaining book.

Consider a lottery in which the odds against your winning millions of dollars are greater than the odds you will be struck by lightning, yet someone is certain to win. It will be almost impossible for the winner not to suppose that some sort of paranormal force smiled on him. Perhaps his winning number was the answer to a prayer. Maybe the number resembles his phone number, his postal code, his birth date, or his social security number. Clearly there is such a multitude of possible correlations that it is not surprising that he may find one. As O'Shea fully understands, there are so many ways that astonishing coincidences can occur in mathematics, science, and other fields as well as in one's life, that if you earnestly search for them you can find them all over the place.

Science is rich in curious coincidences. The moon circles the earth in thirty days, and the sun rotates once in thirty days. Light travels 186,000 miles per second. This is extremely close to the average diameter in miles of the earth's orbit around the sun divided by one thousand. My favorite astronomical example is the almost identical sizes of the disks of the sun and moon when viewed from the earth. The most dramatic consequences of this near identity are the total eclipses of the sun.

Although most science coincidences do not lead to new laws, there are rare occasions when a seeming coincidence turns out, to the surprise of everybody, to lead into significant new knowledge. America's east coast seems to fit the contours of Africa's west coast. For centuries this was thought to be nothing more than an accident, but geologists now know that the two continents were actually joined in the distant past, only later to drift apart. The seeming coincidence that the gravitational mass of a falling body exactly equals its inertial mass played a major role in general relativity. Einstein called it the *principle of equivalence*.

Many elegant mathematical theorems were first suggested by what seemed to be coincidental. A beautiful example is provided by what is known as *Morley's Triangle*. Frank Morley noticed that when he trisected the interior angles of a certain triangle, the three intersections of adjacent triangulating lines marked the corners of a little equilateral triangle. After finding that this was true of any triangle, regardless of its shape, Morley realized he had stumbled on a theorem. Eventually he was able to prove the theorem.

It is impossible to read O'Shea's pages without sensing his great love of mathematics. He makes clear that he is a Platonic realist. Mathematical objects and theorems are discovered, not invented. True, they are abstractions, but they are somehow "out there," with their own mysterious existence, as real, indeed more timelessly real, than the moons of Saturn. Human societies come and go, like all material things, but equations, said Einstein, last forever. In the interior of any sun, Bertrand Russell once wrote, two plus two is four.

To an atheist, G. K. Chesterton remarked, the universe is the most exquisite mechanism ever constructed by nobody. Like Chesterton, O'Shea views the cosmos and its mathematical structure with awe and gratitude. He is impressed by how much we know the behavior of what Einstein liked to call the Old One, but he is equally impressed by the infinite realms about which we know nothing.

Martin Gardner

To My Mother and Father

Contents

Introduction

I write a monthly article on recreational mathematics for the fictional magazine, *The Mathematical Universe*. The magazine circulates in Ireland, the U.K., and, more feebly, in the U.S., Canada, Europe, and elsewhere. However, there are two subscribers in Botswana. It is a magazine produced and written for non-specialists, with a view to giving them information and insight into the wonderful world of mathematics.

Not too long ago, I received a telephone call from the fictional editor of *The Mathematical Universe*.

"O'Shea," he said, "I've got a lead for you."

My fictional editor is a fine and generous man—he has said that someday he might even be able to pay me something for my columns—but he has seen too many old newspaper movies. He would like to be at the center of a busy newsroom, dispatching reporters hither and yon, and although he isn't he tends to talk as if he were. If tobacco were more fashionable than it is, he'd chew on cigar butts.

"What is it?" I asked.

"There's a guy from the States who'll be in Cobh to talk with you. Name of Richard Stein. I've set up a meeting. The Commodore Hotel, 7 o'clock Tuesday. In the bar. Carry a copy of the *Universe* so he'll know you. I told him you'd have one. Give him dinner. Don't send me the bill."

"But...," I started to say. Even though my column was young, less than two years old, I had had experience with correspondents who were wasters of my time, some of them persistent. I had no wish to be trapped with someone who might well be a crank, much less to feed him at my expense.

"Don't argue, O'Shea," my editor said. "He's highly recommended. He'll have some good stuff for your column. It could certainly use it. Talk to him."

Mine not to reason why. I said that I would be at the hotel on time. (I should mention that I am fortunate to live in Cobh, pronounced "cove", County Cork, Ireland, about which I could say a good deal (of which I will give you a small sample in Chapter 8) but this is not a travel book.)

At the appointed time in the appointed place, I was approached by a person who introduced himself as Richard Stein. He was, on the surface, unremarkable. A bit over medium height, a bit under medium weight, hair beginning ever so slightly to thin, soft-spoken, with an American accent. Despite his name, he was, he said, of Irish extraction, "something around three-quarters, plus or minus a few sixteenths," and was going to be in Ireland for an indefinite though limited time. He told me that he had always been interested in recreational mathematics, science, philosophy, conjuring, and all sorts of curiosities and coincidences.

"It makes growing up difficult, as you may know from experience," he said, "when you don't share the tastes of your peers. Early on I was given the derisive title of *The Professor*. Even though I've by now encountered quite a few real professors, I don't mind it. In fact I rather like it."

During dinner I asked him if he was familiar with Ireland or its history. He said he had studied a little Irish history over the years. I found out over time that what was to him a little would be quite a lot to ordinary mortals. He told me that, according to tradition, Ireland's patron saint, Patrick, first came to Ireland in the year 432 A.D. The professor—I'll so refer to him from time to time, because, as readers will see, he certainly deserves some title—pointed out that that was a very suitable date for such a significant event, given the fact that the island of Ireland contains 4 provinces and 32 counties and, what's more, $432 = 4 \cdot 3^3 \cdot 2^2$. He also mentioned that $432 + 1$ and $432 - 1$ (note the 4, 3, 2, 1 sequence) were twin primes. There were more curiosities to follow.

He told me that Ireland consists of 1 major island, 2 jurisdictions, and 4 provinces, that the longitude of the center of Ireland is 8° west of Greenwich, that Ireland had a major insurrection in the year '16 (1916), and the island of Ireland contains 32 counties. The numbers 1, 2, 4, 8, 16 and 32 are the first six numbers in the doubling sequence, and their product, 32,768, is very close to the area of the island of Ireland, 32,588 square miles.

"We are owed 180 square miles by someone," I said. "Where can they be? Did the English take them?"

He said that there may be no missing 180 square miles, because 32,588 is an appropriate area for a country with 32 counties:

$$32{,}588 = 32^3 - (3 \cdot 2)^3 + (3 \cdot 2)^2.$$

The professor also pointed out that the sum of the first six numbers in the doubling sequence, 1, 2, 4, 8, 16 and 32, is 63. He said it was curious that if one uses the simple code $a = 1$, $b = 2$, $c = 3$, and so on, then the sum of the value of the letters in the word *Ireland* is 63, as is the sum of the letters in *Irish*.

Richard told me that he was born in New York on Wednesday, June 6, 1962. (Though I had already surmised it, he also told me that he is a *Richard*, not a *Dick*, a *Rick*, or any other diminutive.) The year was notable not only for his birth, he said, but because $1962/(1 + 9 + 6 + 2) = 1 + 9 \cdot 6 \cdot 2$. His birth date was the 6th day of the 6th month, and $1962 = (6 \cdot 6)^2 + 666$. He said that he would tell me later considerably more about 666, the biblical number of the Beast. Further, he said, 1962 could be written as $1296 + 666$, and the digits of 1296 are the digits of 1962 in a different order. He said he had realized this peculiarity at an early age, and this discovery had increased his natural curiosity about numbers.

The professor mentioned that the Cuban Missile Crisis, which brought the world to the brink of nuclear war, had occurred in October 1962. He said that as a consequence of that crisis 1962 was very nearly the year of the final countdown to nuclear horror. Consequently, he thought it was appropriate that 1962 was part of a countdown: $1962 = 987 + 654 + 321$. It is fortunate, he said, that this is not usually written as $987 + 654 + 321 + 0$, because then the countdown would have been complete.

He said he found it curious that his homeland, which he said he dearly loved, contained 50 states, and that a U.S. 50-cent piece could be changed in exactly 50 ways. (One of the ways included is taking the 50-cent piece and returning another. That's not very helpful for someone wanting change, but it is a fair exchange.) I soon learned that these were just some of the unusual oddities that he liked collecting.

Richard told me that he could still recall how as a young boy he watched the first men walk on the moon. He said that that was a marvelous moment not only for Americans, but also for all humankind. Richard said that President John F. Kennedy had promised the American people in May 1961 that America would do all in its power to land a man on the moon before the end of that decade. When President Kennedy was assassinated in 1963, everything possible was done, Richard said, to ensure that Americans would reach the moon before the end of the sixties. The dream became a reality in 1969 when Neil Armstrong and Edwin "Buzz" Aldrin became the first men to walk on the moon.

Richard said President Kennedy's association with the first moon landing is reflected in a couple of interesting numerical facts. First, Kennedy was born on May 29, 1917. That date, ignoring the year, is usually written in the U.S. as 5/29, or 529. Did I know, asked Richard, that Apollo 11 left for the moon from the Kennedy Space Center in Florida on July 16, 1969 and returned on July 24, 1969? These dates, he said, are usually written as 7/16 and 7/24, or simply as 716 and 724. The professor pointed out that $529 + 716 + 724 = 1969$, the year of the first moon landing.

Richard also asked me to multiply 529 by 716 and divide by 724. I did this on my 10-digit electronic calculator, getting 523.1546961. He then asked me to turn the calculator upside down. When I did so I realized that the first four digits on the left in the display were 1969.

"Before going on," he said, "I'd like to make a statement. Bear with me.

"A natural reaction to that last curiosity, or to any like it, and a reaction that I've had many, many times is, 'So what? Who cares?' That's legitimate. Not everyone can find everything fascinating. But when it comes to mathematics—and my discoveries are part of mathematics, though a humble part—that's too bad. Mathematics is . . . "

He paused.

"Words fail me," he went on, "as they tend to when you encounter something huge, awe-inspiring, and . . . ineffable. Mathematics is the most glorious achievement of the human mind! (Don't interrupt, Owen, even if you don't agree.) It has beauty, depth, it can be delightful and make us laugh with joy, and there is no end to it. It is worth devoting a life to, and it would be a life well spent.

"And what's more, this construction of pure reason gives us power over the world. The planets move in their courses because of mathematics. On a lower level, the bridges we build don't fall down because of mathematics. On a *much* lower level, options traders on stock exchanges get rich because of mathematics. Mathematics has power, at all levels.

"It is also austere, and rigorous, and even unforgiving. You can't fool mathematics. If you make a mistake, you can't cover it up. Degrees of mathematical talent are easy to see. I discovered, long ago, that no matter how much I wanted to I would never be able to prove any theorems that would be worth anything. I don't have enough talent, and there is no way of buying it or faking it. That's the way it goes. I do, though, have a small talent for noticing some things about numbers that other people miss and I've cultivated it, for what it's worth.

"So, Owen, you don't have to say 'So what?' even if that's what you're thinking, though I hope only occasionally. Just stop listening until something interesting comes by."

I assured Richard that I hadn't been bored yet, and was not likely to be. He returned to his topic and said that the first men walked on the moon on July 20, 1969. It was the Apollo program that brought them to the moon. The word

Apollo, he said, contains six letters. The professor pointed out that 6 factorial, usually written as 6!, means $6 \cdot 5 \cdot 4 \cdot 3 \cdot 2 \cdot 1$, or 720. The date the first men walked on the moon was 7/20.

The Apollo 11 astronauts splashed down in the Pacific Ocean, said the professor, at latitude 13° north and longitude 169° west. Consider, he said, the date of the moon landing, 720: $-7 + 20 = 13$. This gives the latitude of splashdown in degrees north. The square of 13, or 169, gives the longitude in degrees west. He showed how that famous date could be reconstructed from two or three copies of itself:

$$196 \cdot 9 + 196 + 9 = 1969$$
$$19 \cdot 6 \cdot 9 - 1 + 969 - 1 - 9 - 6 - 9 = 1969.$$

"I can do that too," I said. "$1969 + 1969 - 1969 = 1969$."

"I'll ignore that," Richard said.

He went on to ask me if I recalled the drama of the U.S. hostages being seized in Iran in late 1979. The professor remarked that the word *Iran* contains four letters, and that the hostages were held for 444 days. Was it not curious, he asked, that $4 \cdot 444 = 1776$, the year the U.S. obtained its independence? He also pointed out that 1776 is 987 plus its reversal, 789, and that 444 is 123 plus its reversal. He startled me when he said that 1776 equals $(177 \cdot 6) + (17 \cdot 7 \cdot 6)$. That famous date can also be expressed with three 1776s as

$$(17 \cdot 76) + (1 + 77 \cdot 6) + (1 + 7 + 7 + 6)$$

and

$$(1 + 77) \cdot 6 + 17 \cdot 76 + 17 - 7 + 6.$$

He then told me that

$$1776 = 12^3 - 4 \cdot 5 + 67 - 8 + 9 = 12^3 - 45 + 6 + 78 + 9.$$

The professor then told me that the representation $1776 = 12 \cdot 3 \cdot 45 + 67 + 89$ is the only way to write 1776 with the nine ascending digits without using exponents. There are, he said, two ways to do the same thing with descending digits:

$$1776 = 9 + 8 \cdot 7 \cdot 6 \cdot 5 + 43 \cdot 2 + 1 = 9 \cdot 87 + 6 \cdot 54 \cdot 3 + 21.$$

There were originally 13 colonies, he said, and $13 = 1^7 \cdot 7 + 6$. There are now 50 states, and $50 = 1 + 7 + 7 \cdot 6$.

The professor also told me of his sincere belief in the power of science to unravel the mysteries of the universe and solve the puzzles posed by nature.

Science consistently produced the goods, he said, and is humankind's only hope of achieving a better future for the entire human race. For this reason, he said, he found it appropriate that the letters of the word *non-scientist* could be rearranged to spell the word *inconsistent*.

I asked Richard if he had ever married, or if he had a family in the U.S. The professor responded by saying that there was a misogynistic Irish joke to the effect that a man never knows what happiness is until he marries, and then it's too late. Later, on a more serious note, Richard explained that he had indeed married when a younger man, but that for a variety of reasons, the marriage did not succeed. He said he now had a partner, Michelle Smit, whom I would later meet. He was vague about what he did for a living, and I did not pursue the subject. Later I found out that he had some connection with the mathematics department at University College, Cork, though not an official one.

As the evening passed Richard explained that he had some information concerning several matters that might be of interest to me. He told me that he had read some of my columns in *The Mathematical Universe*, and found them interesting. The professor told me that he had discovered a number of curiosities on a variety of topics and asked if I would like to hear of them, with a view to having them published in my column.

I was always on the lookout for new material, I said, but I couldn't give any promises about publication, though I would certainly give them the consideration that they deserved. I was afraid that he would try to recycle old material that would be of no interest to my readers.

As you will see in what follows, Richard is no recycler, at least not of numerical curiosities. I hope you will find them as fascinating as I did, and do. As will be described, we met many times, and he gave me much information.

In the course of the evening Richard told me that he had a deep reverence for nature and considered it a great privilege to exist. Even with all the sham and drudgery around us, he said, it is still a wonderful world. The amount of knowledge that the human race had accumulated so far, he maintained, was only a tiny, minute part of the great and wonderful store of knowledge that nature has not yet revealed. Nature, he said, gave up her secrets slowly. They have to be wrestled from her. Then mysteriously, he said, we find that the deepest secrets of nature—when discovered—are the most beautiful. That was a deep puzzle in itself. The professor said he agreed with Edison when he said that "we do not know one millionth of one per cent about anything."

Richard said once again that he was constantly in awe at the sheer beauty, elegance and applicability of mathematics. The mathematician, he said, who discovers a deep mathematical theorem is, in his opinion, looking at beauty bare.

(I recognized the reference to Edna St. Vincent Millay's sonnet, "Euclid Alone Has Looked on Beauty Bare," but I agree with Richard that more people than Euclid alone have had that privilege.)

The mathematician, Richard said, almost invariably describes the various proofs of the deepest theorems in mathematics as beautiful. It is astonishing, he said, that mere human beings, with our very limited intellectual capacity, should find these proofs beautiful and elegant. When a mathematician either proves or is shown the proof of a deep mathematical theorem, he or she cannot help but believe, Richard said, that a great secret of nature has been revealed to him or her.

Albert Einstein, according to Richard, once said in an interview that the most beautiful thing we can experience is the mysterious. Richard said that it was his hope that the curious coincidences, number oddities, and the interesting number properties he was about to convey to my readers would increase their sense of wonder, their sense of awe, their sense of mystery, toward the world. His hope was that by passing some of the information that he possessed on to me it would be brought to a wider audience.

At the end of the evening Richard said that he would like to meet with me over the coming months to discuss some other topics concerning recreational mathematics and curiosities. I told him that I would be delighted to do so. Over the following months the professor gave me some amazing information concerning coincidences and curiosities, usually with a mathematical flavor.

This information will be published in the fictional magazine *The Mathematical Universe* in due course. However, here for the first time, in *The Magic Numbers of the Professor*, you will read some of the fascinating information that Richard Stein has conveyed to me.

There are no proofs of deep mathematical theorems in this book. But scattered throughout its pages are various snippets of information that illustrate the beauty and elegance found in mathematics.

Readers should be warned, though, that the book is not like a novel, to be devoured in one sitting. It is not necessary to rush to the end to see how it all comes out, because there is no plot and no denouement. The book is filled with what I think are delectable items, but consuming too many at one time can lead to premature satiety. You would not (I hope) eat a kilogram of fudge all at once. It would probably make you sick and would certainly put you off fudge for quite some time. So, do not read too many chapters all at once. Take time out, and let your appetite for mathematical curiosities recover in the intervals.

Solutions to some puzzles are given at the ends of chapters. I have also given references for further reading at the end of each chapter. Several of the references are to books by Martin Gardner. *Fifteen* of his books (including *The Magic Numbers of Dr. Matrix*) can now be found on a compact disc, *Martin Gardner's Mathematical Games*, published by the Mathematical Association of America. It is a bargain, and it subsumes many of the references.

I hope that after reading some of the fascinating coincidences and curiosities that occur in this wonderful and amazing world, you will find that your life has been enriched. At the very least that is my wish for you.

Owen O'Shea
Cobh, 2005

CHAPTER 1

Digit curiosities

Richard and I met many times during his stay in Cork, usually over a meal. He invariably let me pay for them. But I did not begrudge the cost because he invariably provided a wealth of material for my column. Sometimes he was accompanied by Michelle Smit, his partner, sometimes not. The chapters that follow record some of our meetings, in no particular order. The chapter headings are only guides for their content, because Richard's mind is very quick and active, the very opposite of ponderous, and tends to flit from subject to subject, sometimes with no transition apparent to an observer. They are there, internally, but Richard feels no need to explain them to slower minds.

May 15, 2004 was a pleasant day in Cork city, as days in Ireland in May go: it was cloudy, of course, but there was no rain, only light winds, and the temperature was predicted to rise to 60°. Though the maximum was actually only 59°, the shortfall was not enough to generate even a single indignant letter to Met Éireann. I had had an appointment with my optometrist with a pleasant result: no glaucoma, no cataracts, and not even a need for new glasses. So it was with a light heart that I made my way to the Hingham Central, an old but refurbished hotel, where I was to meet Richard and Michelle.

He was waiting for me in the Waterside Restaurant. The water consisted of the hotel's swimming pool. When I asked Richard why he had chosen this spot, he said, "Ah, Owen, don't you like the heat and the humidity, and the sound of water splashing? It reminds me of the tropics and the slight whiff of chlorine in the air is reassuring—we are in the *sanitary* tropics."

After renting our space by ordering lunch, we got down to business.

"Here's an item," said Richard. "That $192 + 384 = 576$ is clear. Now, Owen, why do I bring that up?"

Richard, along with altogether too many teachers, has the annoying habit of asking questions to which there is only one correct answer, namely the one that he is thinking of. If you see it, you get no credit for being more than minimally intelligent, and if you don't see it, you have proven yourself inferior once again. You can't win. But I don't begrudge him whatever satisfaction he gets out of it, as long as he doesn't overdo it.

"Ah," I said, "each digit is there exactly once."

"Quite so," Richard said. "What else?"

It is enough to make one gnash one's teeth. Not only do you have to read his mind, you have to read it *completely*.

"Aha!" I said, pleased with my perspicacity. "The digits go up in order on the outside—1, 2, 3, 4, 5, 6—and then continue 7, 8, 9 in the middle, back from right to left."

"True enough, Owen," he said, "but more to the point is that the three numbers are n, $2n$, and $3n$ for $n = 192$. There are only four values of n so that n, $2n$, and $3n$ will contain all nine digits exactly once. Ask your readers if they can find the other three. Here's a multiplication with all nine digits once each: $2 \cdot 6729 = 13458$. Along those lines, ask your readers to find the two numbers that between them use all nine digits from 1 to 9 that have the largest product when multiplied. Just the multiplicands have to contain 1 to 9 exactly once—it doesn't matter what digits appear in the product so long as it is maximal. You might think that $97531 \cdot 8642 = 842862902$ would be the largest, but it's possible to do better. I hope that most of your readers will know what *multiplicand* means, but the decay of the language proceeds posthaste and some may not. Perhaps you should say it differently, though it's a shame when perfectly good and useful words drop out of the language. How will we get along without *subtrahend* and *minuend*? I suppose people will have to point and say 'that thing there' or 'this here.' Such are our uncouth times."

"So they are, alas," I said, though I much prefer the 21st century to the Edwardian age, even the imaginary one that Richard thinks once existed. It's best to agree with people, sometimes, even when you don't.

"Did you ever notice," said Richard, "that the difference between 123456789 and 987654321 is a permutation of the nine digits? It's 864197532. And look at these:

$$291548736 = 8 \cdot 92 \cdot 531 \cdot 746$$
$$124367958 = 627 \cdot 198354 = 9 \cdot 26 \cdot 531487\dot{}$$

"In the first, each of the nine digits appears on both sides. In the second, 124367958 can be expressed as the product of numbers, containing between them all the nine digits, in two different ways. If you like those you will also like this one:

$$335180136^2 = 112345723568978496.$$

Do you see why it is nice?"

"Yes," I said, "it's a square that contains each of the nine digits exactly twice. Is there a square that contains each digit three times?"

"I don't have one with me," Richard said. "But there should be quite a few of them."

He started to think out loud. I didn't follow what he was saying, but I let him go.

"There are $\dfrac{27!}{(3!)^9}$ different 27-digit integers made up of three of each digit," he said. "That's—pass me your calculator, please. I could of course do the estimation in my head but we may as well take advantage of technology while we have it."

He pushed a few buttons.

"Ah, about 10^{21} of them. I wouldn't have thought there were so many. The chance that a 27-digit integer is a square is something around 1 in 10^{13}—I'm being very rough, Owen, I hope you don't mind—so there should be something like $10^{21}/10^{13} = 10^8$ of them. A hundred million! You probably don't want to ask your readers to find one or two. Looking at 27-digit squares can get tedious very quickly."

"There are numerous curiosities with the nine digits," he continued. "For example, $246913578 \cdot 987654312 = 493827156^2$. Speaking of square numbers, the square of 6501 is 42263001 and its reversal, 10036224, is also a square, the square of 3168. Your readers may wish to have the following information for their records. The smallest nine-digital square is $139854276 = 11826^2$, and the largest is $923187456 = 30384^2$. If we call an integer made up of each of 0, 1, ..., 9 just once *pandigital*, and we may as well even though most dictionaries don't include the word yet, then the smallest pandigital square is $1026753849 = 32043^2$ and the largest is $9814072356 = 99066^2$."

"I discovered two curiosities involving all nine digits," said Michelle, "when I was a teenager. I've always liked them, even though the first one is really not surprising if you're not young:

$$987654321 + 1 + 123456789 = 1111111111,$$
$$(18 + 9) \cdot 2^7 = 3456."$$

"The second one is not easy to find at all," said Richard. "Here's another one, where both sides of the equation have all nine digits

$$214358976 = (3 + 6)^2 + (4 + 7)^8 + (5 + 9)^1.$$

It's the only example of its kind, so you don't have to try to find another. Here's another unique equation: the only way to have three fractions containing just the nine digits have a sum of 1 is

$$\frac{9}{12} + \frac{7}{68} + \frac{5}{34} = 1."$$

"That's nice," I said. "Is there anything similar for 2?"

"Despite what people may think, Owen," Richard answered, "I don't know *everything*. Your advanced readers might investigate that and find something new. I should also make it clear that, whatever anyone might think, I haven't discovered everything either. Other people have made contributions. Mike Reid and Philippe Fondanaiche (respectively) found the following two examples of Friedman numbers. A Friedman number, in case you don't know, is a positive integer that can be written in some non-trivial manner using only its own digits, along with addition, subtraction, multiplication, division, powers, and concatenation. A fairly well known example is $2592 = 2^5 \cdot 9^2$. Surprisingly, 123456789 and its reversal are both Friedman numbers:

$$123456789 = \frac{(86 + 2 \cdot 7)^5 - 91}{3^4}$$

$$987654321 = \frac{8 \cdot \left(97 + \frac{6}{2}\right)^5 + 1}{3^4}.$$

The smallest Friedman number is $25 = 5^2$. Your readers might like to find some more."

"Digits are fascinating," I said.

"In the popular mind, certainly," said the professor. "Actually, all integers are equally fascinating, but the single digits have the advantage of familiarity. However, they suffer from being subject to Richard Guy's Law of Small Numbers, that there are not enough small integers to deal with all the tasks assigned to them. Two, for an obvious first example, is badly overworked. There are two parts to the day—daylight and darkness. There are twice two hemispheres in the world—both the Northern and Southern and the sea and land hemispheres. There are two polar regions, Arctic and Antarctic. There are two sexes, which I suppose is a better arrangement than having, say, seven. Each of the twenty-six letters in the alphabet is one of two types—a consonant or a vowel. Computers work on the basis that at any given moment of time an electronic circuit is in one of two states—on or off. There are two main branches of mathematics—pure and applied. There are two main parts of the calculus—differential and integral. All integers are one of two basic types—even or odd. There are two parts of the Bible—the Old and New Testaments."

"Yes," I said, "and there are two conversational states, talking and listening. Can we move beyond dualities, of which there is almost no end?"

"If you like, Owen, if you like," Richard said. He is generally agreeable as long as he can be in the first conversational state. "2 is the first prime, and also the only even prime. Goldbach's Conjecture states that every even number greater than 2 is the sum of two primes. Every Fermat number is the product of all previous Fermat numbers, plus 2."

"What was that?" I asked.

"Fermat numbers, Owen, numbers of the form $2^{2^n} + 1$, $n = 0, 1, 2, \ldots$, or $3, 5, 17, 257, \ldots$. They have the property that $17 = 3 \cdot 5 + 2$, $257 = 3 \cdot 5 \cdot 17 + 2$, and so on. A proof isn't difficult. Carrying on, if p is a prime, then p divides $2^p - 2$. The only prime whose factorial equals itself is 2. Also, $2! + 2 = 2^2$. A good deal less obvious is

$$\frac{11480^2 + 11481^2 + 11482^2 + 11483^2 + 11484^2}{8117^2 + 8118^2 + 8119^2 + 8120^2 + 8121^2} = 2.$$

"That definitely isn't obvious," I said, copying it down. "How did you ever find it?"

"I have my methods, O'Shea," he answered. "Let us move on to 3. Ireland is known throughout the world as a Catholic country. Did you ever realise that the number 3, or its multiples, crop up quite a lot in the Christian tradition?"

"Let me see, I realize that Catholics believe that there are three Divine Persons in the one God. The word *God* contains three letters."

"Correct," said the professor. "There are other instances also. Catholics believe that the Holy Family consisted of three persons, Jesus, Mary, and Joseph. When Jesus was born three wise men went to visit him. It is generally believed by Catholics that Jesus commenced his missionary work when he was thirty and that it lasted for three years. Simon Peter betrayed Jesus three times shortly before he was crucified. Jesus was betrayed for three times ten pieces of silver. He fell three times carrying his cross. Jesus was one of three persons crucified on the one day. According to tradition, Jesus was eleven times three years old when he was crucified. Last, but by no means least, Catholics believe Jesus rose from the dead on the third day.

"Do you remember Otto von Bismarck, Owen?"

"Actually, I do. Prime minister of Prussia, unifier of Germany."

"Very good," said Richard. "Such is the decay of education that the average person in the street doesn't know Bismarck from Adam and might have difficulty locating Germany on a map. You may not have noticed that his name has three parts, that he waged three wars (and signed three peace treaties) and had three children. He died in his 84th, or $3^3 \cdot 3 + 3$ year."

"Some of those may be instances of the Law of Small Numbers," I said. "What about 3 itself?"

"Where to begin?" said Richard, "$3 = 1 + 2$ and $1! + 2!$. It is both the first Mersenne prime and the first Fermat prime. Will you have to tell your readers that Mersenne primes are those of the form $2^p - 1$ where p is a prime, as $2^2 - 1 = 3$, $2^3 - 1 = 7$, $2^5 - 1 = 31$, and so on? I suppose that it's best to take nothing for granted. The first member of the first odd prime pair, 3 and 5, is 3. The triangle, the fundamental shape in geometry, contains three sides and three angles. The physical universe is three-dimensional, at least to casual observers. The trisection of an angle was one of the three famous problems of ancient geometry. The smallest magic square possible is three-by-three. The second triangular number is 3, and every integer is the sum of at most three triangular numbers. I suppose it wouldn't do any harm to show your readers why triangular numbers are triangular."

He scribbled a diagram on a napkin:

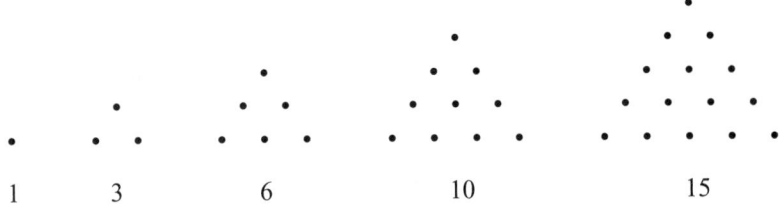

1   3   6   10   15

"The nth triangular number is $n(n + 1)/2$," he said. "You can see that by taking duplicates of those triangles, turning them upside down, and putting them next to the originals. Here's the picture for 6."

"Three rows of four, you see," he said. "Triangular numbers will come up again and again. Tell your readers to be ready. Three primes in arithmetical progression, 3, 5, and 7, start with 3. The sum of the first three positive numbers is the same as the product of the first three positive numbers: $1 + 2 + 3 = 1 \cdot 2 \cdot 3$. At any gathering of three or more persons, three will be either mutual acquaintances or three will be mutual strangers. A number is evenly divisible by 3 if 3 divides the sum of its digits. These are only a few of the places where 3 crops up."

Richard was combining numerical properties of 3 with applications of the Law of Small Numbers, but I didn't point that out. It might have led to a withering reply. It was best to move on, so I said, "What about 11, Richard?"

"Eleven is everywhere," said Richard. "Consider sport. American football, cricket, field hockey, and soccer are all played with teams of eleven players. The sunspot cycle has a period of about eleven years. Also, the number eleven figures prominently in the attempts to get to the moon in the late 1960s. There were eleven manned flights in the Apollo program, two in earth orbit and nine to the moon. The first Apollo flight, Apollo 7, commenced on October eleventh, 1968. It was Apollo 11 that put Armstrong and Aldrin (the first men to walk on the moon) on to the lunar surface. The mission designation of Apollo 11 was AS–506, and $506 = 46 \cdot 11$. Apollo 11 lifted off at 9:32 a.m. on July 16, 1969. (I'll use Eastern Daylight Time, even for times on the moon.) The number of minutes in 9 hours and 32 minutes is 572, or $52 \cdot 11$. It was during the eleventh orbit of the moon that Armstrong and Aldrin crawled through a tunnel for the final check of the lunar module before it separated from Columbia. At eleven minutes past one p.m. on July 20, Armstrong and Aldrin performed a manoeuvre to separate the lunar module from the command service module, and commenced the descent to the moon. Apollo 11 landed on the moon at 4:18 p.m. on July 20, 1969. $418 = 38 \cdot 11$. Neil Armstrong first set foot on the moon at 10:56 p.m. $1056 = 96 \cdot 11$. The Apollo 11 lunar module touched down on a region of the moon named the Sea of Tranquility. The word *Tranquility* contains eleven letters. The first sentence spoken by Armstrong as he set foot on the moon 'That's one small step for man, one giant leap for mankind' contains eleven words. (He did not in fact say 'a man'.) The two astronauts gathered twenty-two kilograms of rocks and soil, or an average of 11 kilograms per astronaut. Shortly after, they re-entered the lunar module and closed the hatch at eleven minutes past one a.m. on July 21. The lunar module lifted off from the moon at 1:54 p.m. July 21, to join the command service module and $154 = 14 \cdot 11$. The command service module was placed in an earth-bound trajectory in the 55th minute of the day on July 22. Both 55 and 22 are multiples of 11. When the astronauts returned to earth the crew recovery helicopter that picked them up in the Pacific Ocean was numbered 66. That number is not only divisible by 11, but it equals the sum of the numbers from 1 to 11. Even the year of the first landing, 1969, is a multiple of 11."

"My," I said. "What else?"

"Nothing more on the moon," Richard said. "But of course there's more down here on earth. The fifth prime is 11, as is the fifth Lucas number. I hope that your readers don't have to be reminded that the Lucas numbers are like the Fibonacci numbers

$$1, 1, 2, 3, 5, 8, 13, 21, \ldots$$

with each the sum of the preceding two, except that we will start the Lucas numbers with 1 and 3:

$$1, 3, 4, 7, 11, 18, 29, \ldots.$$

There are many palindromic primes, but 11 is the only one with an even number of digits. (Every other palindromic number with an even number of digits is divisible by 11. Later I'll give you an easy way to see that.) There is a simple procedure to test any number for divisibility by 11: Add and subtract the digits alternately from either end. If the answer is zero, or a number divisible by 11, the original number is also divisible by 11. For example, take 13574. Starting from the right, we add 4, subtract 7, add 5, subtract 3, and add 1. The answer is zero, so 13574 is a multiple of 11. Take any four consecutive integers greater than 11, and one of them is divisible by a prime greater than 11. Moving back to the heavens, you may be aware that the very last U.K. solar eclipse of the twentieth century occurred in August 1999. The path of the moon's shadow hitting the earth struck Penzance, on the very southwest tip of England, at fifty-six minutes past nine a.m., British Summer Time, on an August morning. The digits of 56 sum to eleven. Mid–totality occurred at Penzance at eleven minutes past eleven a.m. on the morning of August eleventh. That historic eclipse occurred in the 99th year of the century, which, as is abundantly clear, is a multiple of 11. I point that out only because you are looking more than usually dazed, Owen. Shall we stop and do something mundane for a time?"

"No, no," I said. When material for my column is coming thick and fast, it doesn't do to stop the flow. "Please go on."

"I'm reminded," Richard said, "of the coincidences involving John Lennon of the Beatles and the number 9. For example, he was born at 9 Newcastle Road, Penny Lane, Liverpool, on Wednesday, October 9, 1940. (The words *Newcastle*, *Liverpool*, and *Wednesday* each contain nine letters.) The Beatles' first album, 'Please Please Me', went to number 1 in the pop charts on February 9, 1963. Lennon met Yoko Ono for the first time on Wednesday, November 9, 1966. Their son, Sean, was born on October 9. John and Yoko had their last home at the Dakota in New York City, which is on West 72nd Street, in Manhattan. (Note that $7 + 2 = 9$, and that *Manhattan* contains nine letters.) Lennon's hits included 'Revolution 9', '#9 Dream', and 'One After 909'. Lennon was shot on 72nd Street on Monday, December 8, 1980 ($1 + 9 + 8 + 0 = 18$; $1 + 8 = 9$) and declared dead shortly after in Roosevelt Hospital, which is on Ninth Avenue. (*Roosevelt* contains nine letters.) Although he was declared dead on December eighth, in Liverpool, the place of his birth, it was early in the morning of December ninth."

"I don't suppose that there's any connection with the nine planets—Lennon was not a heavenly body. What else about 9?" I asked.

"I could go on for at least nine hours," said Richard. "The difference between 9 and 8 is 1, or $3^2 - 2^3 = 1$. These are the only powers known that differ by 1. (A proof that there are no others was announced in 2002.)"

"What about $17^1 - 4^2 = 1$?" I asked.

"Come, Owen, that is unworthy of you," the professor said. "Non-trivial powers, of course. The only square that is the sum of two consecutive cubes is 9. Also, $1! + 2! + 3! = 9$. All even perfect numbers, except 6, leave a remainder of 1 when divided by 9. The sum of successive powers of 9, plus 1, always yields a triangular number: $9 + 1$, $9^2 + 9 + 1$, $9^3 + 9^2 + 9 + 1$, and so on. ($10 = (4 \cdot 5)/2$ is the sum of the first four integers, $91 = (13 \cdot 14)/2$ of the first thirteen, and $820 = (40 \cdot 41)/2$ of the first forty.) A square whose area is nine units is the smallest square possible on the side of a primitive right-angled triangle. Any number whose digits sum to nine or a multiple of nine is evenly divisible by nine. Nine is the maximum number of cubes required to sum to any number. As water freezes, it expands by nine per cent in volume. Ask someone to write a number down, containing any number of digits, and not to let you see the number. Then give the instruction to scramble the digits of the number, forming a new number and to subtract the smaller number from the larger one. You now know, although you have not seen either number, that the difference is evenly divisible by nine."

"Let me check," I said. "I'll take 1234 and permute it to 3412. $3412 - 1234 = 2178$, sure enough a multiple of 9. Have you any interesting number puzzles that my readers might like?"

"Try these two, Owen. Ask your readers to find two numbers such that the sum of their cubes is a square, and the sum of their fourth powers is a cube."

"That sounds hard, but I'll let them try," I said. "What's the second?"

"Have a look at this," said Richard. The professor wrote some letters and a number on a sheet of paper, and handed it to me. This is what he wrote:

$$4(AB) = CD; CD + EF = GH.$$

"The letters," said Richard, "in that little arithmetical problem represent different digits between 1 and 9 inclusive, and none may be 4. Ask your readers if they can find the solution."

"That sounds difficult as well," I said.

"Life is not easy, Owen," Richard said. "But here is a bet that can make it slightly easier by helping you (and your readers) to collect some loose change. Ask the mark to shuffle a deck of 52 cards and deal them face up. Before he does so, bet the mark even money that of the first king, the first queen, and the first jack that appear, at least two will be of the same suit. If he accepts your bet, it can prove a nice little earner."

"Very nice, Richard," I said. "It seems as if the chance of that happening would be a good deal less than 50%." The correct odds on this bet are explained at the end of this chapter.

"Speaking of cards," Richard said, "do you know the paradox of the second ace?"

"What's that?" I enquired.

"Suppose four people are playing bridge," said Richard. "During the game, suppose that a player states, 'I have an ace.' The chance that he or she holds more than one ace can be calculated. It is 5359/14498, which is less than 37%. Suppose that later the player states, 'I have the ace of spades.' The chance that he or she has more than one ace now is 11686/20825, which is more than 56%. Why should the fact that he or she specified the ace improve the odds?"

"Another question is why a player would make announcements like that," I said. "In a tournament the director would get called in a flash, and I don't know what would happen in a money game. Probably some yelling and screaming, but it's an interesting question. I'll have to work on it." (The reader is invited to figure out why the chance changes when the ace is specified. The solution is given at the end of this chapter.)

"I must ask you Richard, do you have a favorite mathematical puzzle?"

"Ah, Owen, that's like asking if I have a favorite sunset," Richard answered. "It's impossible to rank beauty linearly, 1, 2, 3, ... But here's one I especially like, because the solution involves an elegant shortcut. It appears in the works of both Sam Loyd, America's greatest maker of puzzles, and of Henry E. Dudeney, the great English puzzle genius. Here's one version." The professor drew a right-angled triangle on a sheet of paper.

$a = ?$
$b = 47$
$c = 129$
$d = ?$

What is the distance of a?

"The right triangle represents a piece of ground. All dimensions are in yards. If you walk the distance represented by b and then the distance represented by c, you will have walked 176 ($47 + 129$) yards. If however you walk the distance represented by a and then the distance represented by d, you also walk 176 yards. What is the value of a? The solution applies to all right-angled triangles. To find the value of a, divide c by b and add 2. (One always adds 2.) Then divide this into c. The answer is a. In our example, we would divide 129 by 47, obtaining 2.74468+. Add 2 to this, obtaining 4.74468+. Now divide 129 by 4.74468+, obtaining 27.18834+. Thus the distance of a equals 27.18834. By the Pythagorean theorem we find that d is 148.81165+." Ask your readers if they can figure out why the shortcut works."

"I will," I said.

"The puzzle," said Richard "raises an interesting question. Assume we wish to find a right triangle where the values of a, b, c, and d are all integers. The solution is related to the Fibonacci sequence 1, 1, 2, 3, 5, 8, ... Take any four consecutive terms in the sequence. For example, take 3, 5, 8 and 13. The product of the first and second terms ($3 \cdot 5$) is the value of a. The product of the first and third terms ($3 \cdot 8$) is the value of b. Twice the product of the two inside terms ($5 \cdot 8$) is the value of c. The sum of the squares of the two inside terms ($5^2 + 8^2$) is the value of d. So, in our example, $(a, b, c, d) = (15, 24, 80, 89)$. A nice little bonus is that the hypotenuse d is always a number from the Fibonacci sequence. This is a good example of how two apparently unrelated subjects in mathematics (in this case the right angled triangle and the Fibonacci sequence) are in fact intimately connected. One of the most beautiful and fascinating aspects of mathematics is when one discovers that hitherto apparently unrelated topics in mathematics are in fact intimately related."

"That's very true," I said.

Solutions

1. Readers were told that there are only four values of n so that n, $2n$, and $3n$ contain exactly the nine digits. The four solutions are

n	$2n$	$3n$
192	384	576
219	438	657
273	546	819
327	654	981

2. Readers were asked what two numbers that between them contain all the nine digits once only give the largest product when multiplied. The solution is $87531 \cdot 9642 = 843973902$.

3. The first Friedman numbers after 25 are 121 (11^2), 125 (5^{1+2}), 126 ($6 \cdot 21$), 127 ($-1 + 2^7$), and 128 (2^{8-1}). There are no more consecutive Friedman numbers until ten in a row starting at 2500.

4. Readers were asked to find two numbers such that the sum of their cubes equals a square, and the sum of their fourth powers equals a cube. An example is $289 = 17^2$ and $578 = 2 \cdot 289$. It will be seen that $289^3 + 578^3 = 14739^2$ and $289^4 + 578^4 = 4913^3$.
   These numbers work because the sum of 1^3 and 2^3 is 9, which is a square number, and the sum of 1^4 and 2^4 is 17. So,

$$(289)^3 + (578)^3 = (1 \cdot 17^2)^3 + (2 \cdot 17^2)^3 = (1^3 + 2^3)(17^6)$$
$$= 3^2 \cdot 17^6 = (3 \cdot 17^3)^2$$

and

$$(289)^4 + (578)^4 = (1^4 + 2^4)(289)^4 = 17 \cdot 17^8 = (17^3)^3.$$

   Readers might enjoy searching for similar examples.

5. The solution to that problem where the letters represent eight digits is

$$4(17) = 68; 68 + 25 = 93.$$

6. A deck of cards is shuffled and dealt and the first king, the first queen and the first jack are withdrawn. First, what is the chance that these three cards will be of different suits? The three cards withdrawn can be any one of four kings, four queens and four knaves. Therefore, there are $4 \cdot 4 \cdot 4$ or 64 different possibilities to consider. (The order of the king, queen and jack is

irrelevant.) Suppose the first card to appear is the king. It can be any one of four kings. Suppose the second card is the queen. It can have a different suit from the king in three ways. Suppose the third card withdrawn is a jack. Its suit can differ from both the king and the queen in only two ways.

Thus, there are $4 \cdot 3 \cdot 2$ or 24 different ways that the three cards can be of different suits and so there are $64 - 24$ or 40 ways that at least two of the three cards are of the same suit. The probability that at least two of the cards will be of the same suit is therefore 40/64, or 5/8.

7. To understand the seeming paradox of the second ace, let us first consider a very small deck of cards, consisting of the ace of spades, the ace of hearts, and the 2, 3, and 4 of clubs, and let us deal 2-card hands. There are $\binom{5}{2} = 10$ of them:

$$\text{(AS, AH)} \quad \text{(AS, 2C)} \quad \text{(AS, 3C)} \quad \text{(AS, 4C)}$$
$$\text{(AH, 2C)} \quad \text{(AH, 3C)} \quad \text{(AH, 4C)}$$
$$\text{(2C, 3C)} \quad \text{(2C, 4C)}$$
$$\text{(3C, 4C)}$$

The number of hands with no aces is $\binom{3}{2}$, since we have to choose the two cards in the hand from the three non-aces. The number of hands with exactly one ace is $2\binom{3}{1} = 6$, since the ace can be chosen in two ways and the other card in three ways. So, the number of hands with one ace or no ace is $6 + 3 = 9$ and thus the number of hands with more than one ace is $10 - 9 = 1$. The number of hands with at least one ace is the total number of hands less the number with no aces: $10 - 3 = 7$. If a player has at least one ace (and can say "I have an ace") then the player has one of these seven hands and the chance of having another ace is 1/7.

Now, what if the player says, "I have the ace of spades"? The number of hands with the ace of spades is $\binom{4}{1} = 4$, since that is the number of ways the other card in the hand can be chosen. The number of hands with the ace of spades and no other ace is $\binom{3}{1} = 3$, since that is the number of ways the rest of the hand can be completed without another ace. Thus, the number of hands with the ace of spades and another ace is $4 - 3 = 1$. The chance of holding another ace is therefore 1/4.

Let us repeat the calculation, this time playing with a full deck. There are $\binom{52}{13}$ hands. The number with no aces is $\binom{48}{13}$ and the number with exactly one ace is $4\binom{48}{12}$ (four ways to choose the ace, $\binom{48}{12}$ ways to fill out the hand).

So, the number of hands with more than one ace is

$$\binom{52}{13} - \binom{48}{13} - 4\binom{48}{12}.$$

This corresponds to $10 - 3 - 6$ in the small-deck example. The number of hands with at least one ace is

$$\binom{52}{13} - \binom{48}{13},$$

which corresponds to $10 - 3$ in the small-deck example. So, the chance of a hand having more than one ace given that it has one ace is

$$\frac{\binom{52}{13} - \binom{48}{13} - 4\binom{48}{12}}{\binom{52}{13} - \binom{48}{13}},$$

which, if you take the trouble to compute it, is about .396.

The number of hands with the ace of spades is $\binom{51}{12}$, and the number with the ace of spades and no other one is $\binom{48}{12}$. So, the number with the ace of spades and at least one other ace is

$$\binom{51}{12} - \binom{48}{12}.$$

Thus the chance of having another ace, given that the hand has the ace of spades, is

$$\frac{\binom{51}{12} - \binom{48}{12}}{\binom{51}{12}},$$

which is approximately .561. Richard said that this could be gotten more quickly by using conditional probabilities and that readers who knew about them should try it.

8. In the triangle problem we were asked to explain why the rule works that so neatly gives the value we are seeking From the conditions of the problem we know that $a + d = b + c$, or $d = b + c - a$. Squaring,

$$d^2 = b^2 + c^2 + a^2 + 2bc - 2ab - 2ac.$$

We know from the Pythagorean theorem that $(a + b)^2 + c^2 = d^2$. Thus

$$d^2 = a^2 + b^2 + c^2 + 2ab.$$

Comparing, we see that

$$2ab = 2bc - 2ab - 2ac$$

or

$$2bc = 4ab + 2ac.$$

That is,

$$bc = a(2b + c),$$

from which we get

$$a = \frac{bc}{2b + c} = \frac{c}{2 + \dfrac{c}{b}}.$$

The last equation above tells us that if we divide c by b, add 2 to the result, and then divide that result into c, our answer will be a, whence the shortcut in solving the puzzle. The problem appears in *More Mathematical Puzzles of Sam Loyd*, (pages 67 and 151) published by Dover, 1960, and edited by Martin Gardner.

References for further reading

Martin Gardner, *Mathematical Puzzles and Diversions*. Penguin Books, 1965, chapter 5.

John Haigh, *Taking Chances*. Oxford University Press, 1999.

CHAPTER **2**

The 9/11 atrocities

Richard was back from Dublin.

"A successful trip, Owen," he said. "I was able to locate a copy of the *Dublin University Magazine* for 1842, the one with a biographical sketch of William Rowan Hamilton written *before* he discovered quaternions."

"Should I tell my readers about quaternions?" I asked.

"Probably not," he said. "The field of recreational quaternions is undeveloped and likely to stay that way."

He paused, looking into the air.

"Hmm. . . ." he said. "There's. . . ." Another pause. "I'll have to think about that, but later."

I may have witnessed the birth of recreational quaternions.

"You should remind your readers about Hamilton, though," he said. "He was Ireland's greatest mathematician, after all. He had an extraordinary mind. See what I found in the *Magazine*, written by somebody named Robert Graves. I love early Victorian prose:

> In consequence of Mr. A. Hamilton, the father, having some friends among the body who then held the patronage of India, he originally destined his son to a life in the East, and accordingly directed that the mind of the child should be early employed in the acquisition of the oriental languages. Happily the subsequent development of his scientific powers frustrated this plan, but its immediate results were too remarkable in themselves, and for the proof they give of the activity and versatility of his faculties, to allow us to pass them unnoticed. At the age of four he had made some progress in Hebrew: in the two succeeding years he had acquired the elements of Greek and Latin; and when thirteen years old was in different degrees acquainted with thirteen languages, besides the vernacular—Syriac, Persian, Arabic, Sanskrit, Hindustani, Malay, French, Italian, Spanish and German; and we are not sure that this list is a complete one. We well remember to have heard, long before we ever saw our friend, of Dr. Meredith, formerly

fellow of Trinity College, and a man of great learning and ability, reporting with expressions of astonishment, that he had examined in the country a child of six or seven, who read and translated and understood Hebrew better than many candidates for fellowship; this child was young Hamilton: we know also that he not infrequently wrote letters in Persian; and we think the anecdote should not be lost, that one which he sent in that language as a greeting to the Persian Ambassador, Mirza Abou Hassan Khan, when on a visit to Dublin in 1819, drew from the ambassador the exclamation, that he did not think there was a man in these countries who could have indited such a letter.

"Isn't that amazing?" he said. "It's greatly to the credit of the Irish that they haven't forgotten Hamilton. Give me €32, Owen."

"What for?" I asked. It was an abrupt transition.

"So I can get a €10 coin," he answered.

"That doesn't sound like much of a bargain," I said.

"Ah, but it's not just any old €10 coin," he said. "It's a special Hamilton commemorative coin. They're going to make only 30,000 of them. In fact, give me €64 and I'll get one for you. You won't regret it. It'll be a great investment."

"I'll think about it," I said. "While I'm thinking, do you have any good items for my column?"

Richard told me that he had some curiosities concerning the dreadful 9/11 attacks, and asked if I would like to use them in my monthly column in *The Mathematical Universe*. I responded by saying that I was always on the lookout for new material, and would certainly consider anything that he had to offer me.

"The 9/11 attacks were horrible," I said. "The world will never be the same again."

"They were absolutely dreadful," said the professor. "A terrible example of man's inhumanity to man."

"I have no wish," I said, "to discuss the dreadful carnage and suffering caused by the attacks. But you say you have some curiosities of a mathematical nature relating to the attacks that my readers might find interesting?"

"Yes, I have, Owen. It is only to be expected that there should be several number curiosities associated with this terrible event."

"I do know of a number of curiosities concerning the 9/11 attacks," I said. "Perhaps I will state these so you can then tell me about curiosities that I do *not* know about."

"OK. That sounds good."

"Well," I said, "First, the date of the attacks, 9/11, is curious, given that the emergency phone number in the US is 911. I am also aware of the appearance of the number 11 in various aspects of the attack. The Twin Towers themselves looked like a large number 11 on the New York skyline. There are 11 letters in the words *New York City*. The plane that smashed into the North Tower was American Airlines Flight 11. (The plane that hit the Pentagon incidentally, was AA Flight 77, seven times eleven.) The South Tower burned for 57 minutes, then it collapsed in 11 seconds.

"That's correct," said Richard.

"There's more," I said. "The attacks happened on the 254th day of the year. The sum of these digits is 11. The suspected leader of the group behind the attack is Usama bin Laden. He was said to be based in Afghanistan. The name of that country contains 11 letters. Its capital is Kabul. Its initial letter is the 11th letter of the alphabet."

"Yes," said Richard, "I know that."

"Ruth McCourt from Cork, here in Ireland," I said, "and her young daughter were both killed on Flight 175 when it smashed into the South Tower. By a remarkable coincidence, Ruth's brother, Ronnie Clifford, was, at the time of her death, helping to pull victims from the debris of the North Tower over one thousand feet below. He looked up and saw Flight 175 slam in to the South Tower, not knowing that his sister and niece were on the plane."

"Yes," said Richard, "I read of that. It was dreadful."

"The attacks were inhuman," I said." Words cannot adequately describe the horror of this wicked deed."

"I agree entirely, Owen," said Richard. "However, without wishing in any way to diminish the horror of the attacks, and fully recognising the terrible suffering caused by these inhuman acts, I would like to give you some curiosities about 9/11 that have never before been published. For example, here is how the number of the Beast was involved:

7	days in a week
52	weeks in a year
365	days in a year
666	number of the Beast
911	date of the attacks
2001	sum—the year of the attacks.

Consider a Fibonacci sequence, not the usual one starting with two 1s, but one starting with 9, 11. Go ahead, write it out, to the 11th term."

I did:

$$9, 11, 20, 31, 51, 82, 133, 215, 348, 563, 911.$$

"Amazing," I said. "Is there any other connection between the 9/11 attack and 666, the number of the Beast?"

"It appears in $91.1^{0.2001}$, which is $2.466605\ldots$ Incidentally, the date of the attacks, 9/11, in year 1 of the century, may be written as 9/11/1, or simply as 9111. The natural logarithm of 9111 is $9.11\ldots$"

Just to be doing something, I checked this on my calculator. Of course, the professor was correct.

"I know that there were 110 floors in each of the Twin Towers," I said. "Is there any thing numerically interesting about that?"

"Need you ask, Owen?" he answered. "The attacks happened on the 11th day of the 9th month and $\frac{11!}{9!} = 110$. Speaking of floors, the first plane crashed into the 96th floor of the North Tower. Note that $\frac{96^2}{911^2} = 0.0111\ldots$ and the 9/11 attacks occurred 111 days before the end of the year. Further, $\frac{96^2}{911} = 10.116\ldots$ Turn that upside down and reverse it and there is 9 11 01, the date of the attacks."

"Are the numbers trying to tell us something?"

"Some people may think so," said Richard. "One thing they are telling us is that numbers are glorious and inexhaustible, which may be a sufficient message."

"Have you any number coincidences concerning the attack on the Pentagon?" I asked.

"Certainly. Believe it or not, construction work on the Pentagon commenced on Thursday, September 11, 1941, sixty years to the day before the attacks. The Pentagon is, of course, a five-sided building. It has five rows of offices at each of its sides. Including the basement, it has five floors. There are three fives there. Now $5 \cdot 5 \cdot 5 = 125$. There were 125 people killed (inside the Pentagon) in the attack on the Pentagon. Consider the two words *the Pentagon*. Use the code $a = 1$, $b = 2$, $c = 3$, and so on. Substitute the appropriate values for the letters in the words *the Pentagon*, and sum the values. The result is 125. Given the fact that construction work on the Pentagon commenced in 1941, you may find the following equations interesting:

$$1941 = 911 + 911 + 119 \text{ (the last number is the reversal of 911)}$$
$$911 + 911 + (9 \cdot 11) + 9 + 11 = 1941$$
$$19^2 + 41^2 - 41 = 2001, \text{ the year of the attacks}$$
$$19 \cdot 41 + 41 + 91 \cdot 1 = 911$$
$$2001/1941 = 1.03091190\ldots \text{ (note the 911)}.$$

As a bonus, in the last equation the first seven digits give $1030 + 911 = 1941$, the year the Pentagon was built. Ask your readers if they can arrange the digits 191, 911, and 119, in that order, in an equation using only the plus, minus, and multiplication signs and parentheses, so that it equals 2001. They can insert the plus, minus and multiplication signs of arithmetic anywhere, and parentheses also. It isn't a difficult puzzle."

"I'll give it to my readers," I said. "I'm sure they will enjoy it."

Richard then pulled out a twenty-dollar bill and folded it in a curious manner. One side of the bill depicted the Twin Towers ablaze after the planes struck. (See Figure 1.)

FIGURE 1

The other side depicted the Pentagon on fire after it was attacked. (See Figure 2.)

FIGURE 2

The professor then crumpled the $20 bill in his hand and gave it to me.

"Look at the name that appears on the bill," he said. He handed me the note. (See Figure 3.)

FIGURE 3

I was astonished. There, on the bill was the name, *Osama*.

"There was a strange coincidence in the New York Lottery on September 11, 2002, one year exactly after the attacks," Richard said.

"I missed that, Richard. What happened?"

"The winning numbers in the evening draw on that day were 9, 1, 1. There was just one chance in a thousand that those particular numbers would come up. 911 has three permutations, 911, 191, and 119. The 9/11 attacks occurred in year 1 of the century. This gives us four numbers, 911, 191, 119, and 1, whose sum is 1222. In base 9, 1222 is 911. Take another representation of the date of the attacks: 9/11, 2001. You probably have not noticed that $9.11 + 2.001 = 11.111$. 11 signifies that the attacks occurred on the 11th day of the month and the 111 at the end signifies that the attacks occurred 111 days before the end of the year. What's more, $11 \cdot 111 = 191 + 911 + 119$, and the first three digits of $191^{0.911}$ are 119."

"You amaze me, Richard," I said. "Surely that's everything."

"Not at all," he said. "Large events have a large number of consequences. For instance, 9/11 is the 254th day of the year—see 911 appear straddling the decimal point in $(1941 \cdot 2001)/254 = 15291.106\ldots$. The following equation is curious:

$$9^{(1+1)} + 9^{(1+1)} + 91 + 1 = 254.$$

9/11 expressed in Roman numerals is IX/XI, which is a palindrome. Note the two 1s at each end of the expression give the day of the month of the attack and the two crosses inside signify that a major double cross occurred on that day. The digits of 911 are connected with the previous attack on the Twin Towers. It occurred on Friday, February 26, 1993. The period of time between the two attacks is 3119 days. Note the appearance of 119. The number of days between September 11, 1941, the day the first stone was laid in the building of the Pentagon, and September 11, 2001, is 21915. Note the appearance of 191. The Pentagon was built in 1941, a year that contains the three digits of the number 911. The digit 4 in the number represents the fact that four planes were involved in the attacks. The attacks occurred in 2001, on the 3rd day of the week, the 9th month of the year and the 11th day of the month. Write this as 20013911 and partition it as 20013, 911. Then $911^{0.20013} = 3.911\ldots$. The attacks occurred on the 3rd day of the week, in the 9th month, on the 11th day. The Pentagon was built in '41. The number of the Beast is 666. Multiply: $3 \cdot 9 \cdot 11 \cdot 41 \cdot 666 = 8109882$. The difference between the date of the attacks, 9112001, and its reversal, 1002119, is 8109882. By the way, did you know that 191 in base 26 and in base -35 is 911?"

"Base -35?" I asked. "That's an odd way to count."

"Yes," Richard said, "we use base 10 because we have ten digits on our hands, and nothing has -35 digits. Nevertheless,

$$1 \cdot (-35)^0 + 9 \cdot (-35)^1 + 1 \cdot (-35)^2 = 1 - 315 + 1225 = 911.$$

"The number 119 can be partitioned as 1, 19. The first 1 signifies the one hijacker who (reportedly) did not turn up on September 11, 2001. The 19 represents the number of hijackers who, unfortunately, did turn up. It is also curious that 9112001 is divisible by 19^2. The sum of the numbers from 1 to 19 is 190. Add 1 to this (to represent the one hijacker who did not turn up on the day) and we have 191. Write 'nine eleven two thousand and one'. Under each word write the number of letters in that particular word and the following arithmetical signs, like this."

The professor took out his notebook and wrote

nine		eleven	two		thousand	and		one
(4	+	6)	+ (3	+	8)	+ (3	·	3)
	10			11			9	

"Do you see the significance?" he asked.

"Not quite," I said. The flood of numbers had numbed my mind.

"It's the reversal of the date of the attack, 9/11/01," he explained.

"Of course it is," I said. "It's obvious when you point it out."

"The ability to see the obvious is to be treasured. You must have heard the anecdote about the mathematician (several different mathematicians in different versions of the story) who, while lecturing, said, 'It is obvious that...' and then paused. In some versions he left the room, in others he wrote on a sheet, or sheets, of paper, and in some he merely thought very hard for a length of time, but the versions all agree that, after the hiatus, he went on, 'Yes, it is obvious that....' and continued his talk. I expect, Owen, that you don't know what day it is."

"It's obvious," I said, "it's right here on my watch."

"Not *that* day," Richard answered, "the Julian Day."

"The what day?" I asked.

"The Julian Day Number," the professor said, slipping easily into lecture mode, "is a system of dates used by astronomers when working with different calendars to unify different historical chronologies. The Julian Day Number is the number of days that have elapsed since 12 noon, Universal Time (formerly called Greenwich Mean Time—the name is one more thing the British lost with their empire) on Monday, January 1, 4713 B.C. in the proleptic calendar."

"The what calendar?" I asked, as I think Richard expected me to do.

"Proleptic, Owen," the professor said. "From the Greek. It's the calendar extended backward. You don't think our calendar was in effect in 4713 B.C., I hope."

I didn't answer that.

"Each day in the Julian Day system begins and ends at noon. The Julian Day beginning at noon, January 1, 2001 was number 2,451,911. It contains 9/11, you'll notice. Also,

$$2 \cdot 451 + 9 + 1 - 1 = 911.$$

Yet further, $\sqrt{2451911} = 1565.8579118....$ There 9/11 is again. Partition the number as 245 1 911. The central 1 shows that a significant event would occur in the first year of the new millennium. As for the outside numbers, $911 - 245 = 666$, the number of the Beast."

"Perhaps watches should be made to show the Julian day," I said, before changing the subject. "Is there any numerical connection between the 9/11 attacks and other world famous events, such as the assassination of John F. Kennedy?"

"Indeed there are," said the professor. "Catastrophe, as well as misery, loves company. Kennedy was known by his initials, JFK. Those are the tenth, sixth and eleventh letters of the alphabet. Look at 10 6 11 upside down and read it backwards and you have the date of the attacks as dates are customarily written in Europe."

"George W. Bush would be 7 23 2, which can't be turned upside down, so he isn't telling us much" I observed.

"Most presidents don't. Besides Kennedy, only John Adams, Andrew Jackson, and Andrew Johnson have had initials that can be representable by upside-downable digits."

"What about the number 911 itself?" I asked. "Does it crop up in any interesting equations?"

"Certainly," he said. "Here's one with all the digits from 1 to 9, with first the even digits in order and then the odd digits in order:

$$2 \cdot 468 - 1 - 3 - 5 - 7 - 9 = 911.$$

Here are two other representations of 911, each with three 911s:

$$(9^{1+1})(9 + 1 + 1) + (9 + 11)$$
$$(9 + 1 - 1)(9 \cdot 11) + (9 + 11).$$

Incidentally, consider the date of the attacks written as 9/11, 2001 and check the square root of 9112001."

I did: it's 3018.60911.... "There's 911 again," I said.

"Yes," said Richard. "The number of dead resulting from the attacks is calculated as follows: the number killed in the Pentagon is 125, with an additional 59 dead on the hijacked plane. In Pennsylvania, 40 people died on the hijacked plane. The number killed in the attacks on the World Trade Center (these figures were published by U.S. authorities on March 19, 2004) is said to be 2,749. The number of dead in the 9/11 attacks is thus 2,973. (This does not include the 19 hijackers killed in the attacks.) The attacks occurred on the 3rd day of the week. Is it not interesting that

$$3(911 + 91 \cdot 1 - 9 - 1 - 1) = 2973?$$

Here is another weird statistic. The total number of unidentified dead at the World Trade Center is 1222, which is $911 + 191 + 119 + 1$. The first three numbers are the three permutations of 911. The number 1 signifies that the attacks occurred in the first year of the century. The total number of dead (including the 19 hijackers) is 2992 which multiplied by 911 is 2725712. It is curious that $272 + 5712$ is twice 2992. Your readers may be interested to know that the sequence 9112001 first appears in the decimal expansion of π at position 36916577, counting from the first digit after the decimal. Note that

$$3^6 - 9(1 + 6) + 5 \cdot 7 \cdot 7 = 911."$$

"Do you recall the old puzzle," I said, "about putting plus or minus signs wherever you please inside the ascending series 123456789 and also in the reverse descending series 987654321 to get 100?"

"That's a very old chestnut," Richard said. "There are many solutions.

$$98 + 7 - 6 + 5 - 4 + 3 - 2 - 1 = 100$$

comes to mind."

"Is it possible to insert the plus and minus signs only into the series 123456789 or the series 987654321 so that the series totals 911?" I asked.

"I was thinking about that problem myself," said Richard. "A good friend of mine, Dr. Joe Manning, a computer scientist in University College, Cork, wrote a computer program to check this. He found that there are no solutions if you insert only plus or minus signs in either the ascending or descending series."

"Ah, well, such is life."

"It is indeed usually mundane and filled with disappointments," said Richard. "However, if you can insert the plus, minus and the multiplication sign into either of the two series, there are four solutions for the ascending series, and eleven solutions for the descending series. There is just one solution for the ascending series that uses single digits. The same solution in reverse order is, of course, the one solution for the descending series that uses single digits."

Richard furnished me with the complete set of solutions, for both the ascending and descending series. I give them in the answer section.

"If you can place a minus sign in front of the first digit, there is just one solution for the ascending series, and only three solutions for the descending series. Your readers may enjoy searching for them.

"It is curious," said Richard, "that the first major international terrorist attack following the 9/11 atrocities—the bomb attack in Bali, in Indonesia on October 12, 2002—has a numerical association with the 9/11 attacks."

"What is that?" I asked.

"The number of people killed in the Bali attack has been officially given as 202, and $202 = 9^2 + 11^2$. Also, $911^2 - 202^2 = 789117$—another 911. The dates of the two attacks," said Richard, "are in arithmetic progression. The attack on America was on 9/11/01. The attack on Indonesia was on 10/12/02. We're fortunate that the pattern didn't continue on 11/13/03 or 12/14/04.

"The 9/11 attacks have had far-reaching consequences. In November 2002, the United Nations passed Resolution 1441 in relation to Iraq. The product of 911 and 2001 is 1822911 and the natural logarithm of 1822911 is 14.41...."

I asked the professor if he had any observations on the terror bombing of Madrid on March 11, 2004. Initial reports said that almost two hundred people were killed in the terror attack. (According to a CNN News report on April 5, 2004, the number of people killed in the Madrid train bombings was 191.)

"You should ask me about happier events, Owen," he said. "They can have connections to other happy events, even as disasters are connected to other disasters. The period of time between the U.S. attacks on 9/11 and the Spain terror attack is exactly 911 days. The attack in Spain occurred on the 71st day of the year, in the third month. Look at the last three digits in $71^3 = 357911$. That number, incidentally, contains five consecutive odd numbers, in ascending sequence, beginning with 3. Here is one last elegant representation of 911:

$$1 - 23 + 4! + 5! + 6! + 78 - 9 = 911.$$

And here are some opportunities for your readers. Take the date of the attack,

$$9 \quad 11 \quad 20 \quad 01,$$

as the first row of a magic square. Complete it so that all rows, columns, and the two main diagonals sum to 41. It's not a requirement to use consecutive integers. In the example that I found, there is no 10, for example, though 1, 2, . . . , 9 all appear. Of course, there should be no duplication of entries, and no negative numbers."

"That may be hard for my readers," I said. "Some of them will get it, though."

"Here are two other curiosities that may be presented as little puzzles. It appears that all nineteen hijackers—all male—were in the USA for quite a period of time prior to the attacks. Consider the sentence *Male Band in USA*. Rearrange the letters of this sentence to find the name of an individual central to the events of 9/11. Next, write the name *Usama Bin Laden* using the code $a = 1$, $b = 2$, $c = 3$, and so on. (Transliteration from Arabic to English may be done in more than one way, and *Usama* is a variation of *Osama* that could be used.) Sum the numbers obtained, and turn the answer upside down. You will probably find the answer surprising."

I quickly did what the professor asked, and was pleasantly surprised.

"Have you any other interesting puzzles I can give my readers that may be associated with the 9/11 attacks?"

"Here's one your readers might appreciate." The professor then gave me the following problem.

Consider the pentagon in Figure 4. Each side of the pentagon is one unit

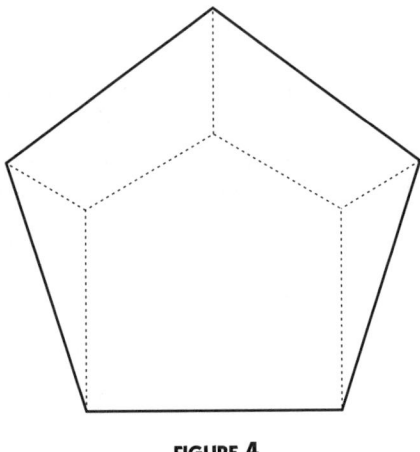

FIGURE 4

long. The seven broken lines drawn inside the pentagon are such that the total length of the seven lines is the minimal length that will join all five vertices of the pentagon. The dotted angles inside the pentagon are 120 degrees. What is the total minimal length of those seven lines? The reader will, I hope, find the puzzle interesting, but will also find that the answer contains an unexpected but pleasant surprise.

"One final question, Richard. Why do you think the date 9/11 was chosen for the attacks?"

"I have no idea," said Richard. "But if I wished to choose a date for an event of worldwide significance to occur, I may have proceeded as follows: Consider the number of days in an ordinary year, 365. I would sum the first and second digits of 365, and then sum the second and third digits. That would give 9/11."

Solutions

1. The reader was asked to arrange the digits 191, 911 and 119, in that order, and to insert only parentheses and the plus, minus, and multiplication signs to represent 2001. It is likely that the only solution is $(-1 \cdot 9) + 1911 + (11 \cdot 9) = 2001$.

2. Here are all the ways of representing 911 using the nine digits in ascending and descending order, and only addition, subtraction, and multiplication:

$$1 \cdot 2 - 3 + 4 \cdot 5 \cdot 6 \cdot 7 + 8 \cdot 9 \quad \text{(uses only single digits)}$$
$$12 \cdot 34 + 5 - 6 + 7 \cdot 8 \cdot 9$$
$$1 \cdot 23 \cdot 4 + 5 \cdot 6 + 789$$
$$1 - 2 + 3 + 4 \cdot 5 \cdot 6 + 789$$

$$9 + 876 + 5 \cdot 4 + 3 + 2 + 1$$
$$9 \cdot 87 - 6 + 5 + 4 \cdot 32 + 1$$
$$9 + 8 + 765 + 4 \cdot 32 + 1$$
$$98 + 76 \cdot 5 + 432 + 1$$
$$987 - 65 - 4 \cdot 3 + 2 - 1$$
$$987 - 6 \cdot 5 - 43 - 2 - 1$$
$$987 - 65 - 4 - 3 \cdot 2 - 1$$
$$987 + 6 + 5 - 43 \cdot 2 - 1$$
$$9 \cdot 87 + 6 - 5 + 4 \cdot 32 - 1$$
$$9 \cdot 8 + 7 \cdot 6 \cdot 5 \cdot 4 - 3 + 2 \cdot 1 \quad \text{(uses only single digits)}$$
$$9 + 876 + 5 \cdot 4 + 3 \cdot 2 \cdot 1.$$

Allowing a minus sign in front of the first digit, there are four more solutions:

$$-12 - 3 + 4 \cdot 56 + 78 \cdot 9$$
$$-9 + 876 + 5 \cdot 4 + 3 + 21$$
$$-9 + 8 \cdot 76 - 5 - 4 + 321$$
$$-98 - 76 + 543 \cdot 2 - 1.$$

3. A solution to the professor's magic square puzzle is

9	11	20	01
5	6	4	26
8	17	14	2
19	7	3	12

4. The letters in the phrase *Male band in USA* can be rearranged to spell *Usama bin Laden*. They can also be rearranged to form *A bad male in sun*, appropriate for Saudi Arabia, where it is usually cloudless. The alternate spelling *Osama bin Laden* yields *Is a lone bad man*.

5. Using the code $a = 1, b = 2, c = 3 \ldots$, the letters of *Usama bin Laden* sum to 116, or 911 upside down and reversed. Using the same code, *New York* has value 111. 9/11 was 111 days before the end of the year.

6. The minimal length is 3.89115. . . . Note the 911.

References for reading

A Nation Challenged. A Visual History of 9/11 and its Aftermath. Jonathan Cape, nd

CHAPTER 3

The professor speaks on the U.S. and Ireland

Talking about the weather is always a good and safe way to start a conversation.

"So, Richard," I said, "how do you like our May weather? That 30-mile-an-hour wind seems to be turning people's umbrellas inside out. I heard that the barometric pressure is below 29 inches and still falling. Wretched! I don't suppose it's because 29 is a sum of two squares and 30 isn't?"

"I hadn't really noticed," Richard said, even though he was dripping on the floor. "About the weather, that is. Anyone would notice about sums of two squares, of course. The next integer along, 31, isn't, but 32 is. Then 33 isn't, but 34 is. It's tempting to go on, but I'll resist. I have some weather-related puzzles that might give your readers a little diversion. Here's an old one: if it is zero degrees Celsius today, and it was twice as cold yesterday, what was the temperature yesterday?"

"I think I know the answer to that one," I said, "but I'll give it to my readers. Any more?"

"Try this one," said Richard. "If n degrees Fahrenheit is the exact same temperature as n degrees Celsius, what is the value of n?"

The solutions to the two puzzles are given at the end of the chapter.

"That has been noticed many times," Richard went on. "It's not as well known that 527 degrees Fahrenheit can be converted to Celsius by moving the first digit to the end, obtaining 275 degrees Celsius. What is the next Fahrenheit temperature with this property?"

"I can never remember the formula—is it $F = \dfrac{5}{9}C + 32$ or $F = \dfrac{9}{5}(C - 32)$? I'd have to look it up"

"Actually, Owen, if you remember that $0°$ C is $32°$ F, $100°$ C is $212°$ F, and the scales are linearly related, you could derive the formula in a moment."

Richard really does border on the other-worldly sometimes. Derive a formula! "Well, my readers will need the formula. What is it?"

Richard sighed. "It's $F = \dfrac{9}{5}C + 32$ or, the other way around, $C = \dfrac{5}{9}(F - 32)$. The problem may be too much for them, formula or not. The next example is $5294117647058823527°\ F = 2941176470588235275°\ C$."

"That's really hot!" I said. "Can you ever convert Celsius to Fahrenheit by moving a digit to the end?"

"That," said the professor, "is a topic for investigation."

I can never tell if Richard knows the answer and is not letting it out, or whether he is as ignorant as I. I had some time ago decided that it was best not to press him.

I had earlier mentioned to Richard that I intended writing an article on coincidences for my column in *The Mathematical Universe*, and would appreciate any assistance that he could give me.

"What is your opinion on coincidences?" I asked. "Do you think that sometimes there are supernatural forces at work when astonishing coincidences occur?"

"Not at all," said Richard. "There are over six billion people on this planet. Most of them are doing many hundreds of different things each day. From that perspective it would be strange if seemingly strange things did not happen."

"I agree," I said. "Nevertheless, one cannot but be startled when one encounters a truly strange coincidence."

"True," said Richard. "It seems that our minds become attuned to the mundane. Our intellect then responds differently when we learn of, or indeed, experience an unusual coincidence. Consider, for example, *The Narrative of Arthur Gordon Pym* by Edgar Allan Poe in which there is a shipwreck where three sailors and a cabin boy survived in a small open boat with a very limited supply of food and water. Their supplies eventually ran low and the three sailors conspired amongst themselves to kill and eat the cabin boy, whose name in Poe's story was Richard Parker. Nearly fifty years later, in real life, there was a shipwreck, in which three sailors and a cabin boy survived in a small open boat, with very limited supplies of food and water. Eventually, the three sailors killed and ate the cabin boy, whose name was also Richard Parker."

"That's amazing!" I said.

"No, Owen, it only *seems* to be amazing," Richard said. "Think of all the millions of incidents described in all the stories and novels ever written. Think of all the incidents that actually occur—the number must be in the trillions at least. The probability is essentially 1 that matches will occur. Curiosities crop up all over the place, even in other books. In 1972, *Electronics for Schools* was published. Its author was R. A. Sparkes. In the same year there appeared *The Imperial Animal*, by Lionel Tiger and Robin Fox. There are other coincidences that are not so well known, but that are interesting nevertheless. Consider what happened on Wednesday, November 14, 1973, the day that Princess Anne and Captain Mark Phillips were married at Westminster Abbey in London. At the Wolverhampton races that afternoon the Royal Wedding Handicap Chase was run at 3:15 p.m. The horse that won that race was named Royal Mark. Another curious coincidence concerns the three major amateur sports played in Ireland, hurling, Gaelic football, and camogie."

"What, Richard, you know about our national games?" I said.

"Certainly," Richard said. "One must participate in one's time, and in one's place. Don't display Irish chauvinism, Owen. Even an American can know that in camogie the hurley may be dropped to handpass the sliotar, while that's a foul in hurling. I'm sure that were you to come to the U.S. you would soon master baseball's infield fly rule. But, as you know, an All-Ireland Final in each game is played annually. In 2003, the third year of the third millennium, the winning margin in all three finals was just three points. All three finals, incidentally, were played in the ninth (3^2) month.

"The Celtic Football Club in Glasgow," Richard went on, "has what might seem to be altogether too many connections with 25 and its reversal, 52, if you were not aware of the power of coincidence. The club was formed in 1888, whose digits sum to 25 and $1888 = 2^5 \cdot 52 + 2^5 \cdot (5 + 2)$. What's more, $1888 = 2^5 \cdot (2 \cdot 5 \cdot 5 + 2 + 2 + 5)$. On May 28, 1888 the club played its first match and won, 5-2. Celtic beat Queen's Park 5-2 to win the Scottish Cup for the first time in 1892. Their most famous manager, Jock Stein (no relation, as far as I know), became captain of the club in 1952. When he left it in 1978 (whose digits also sum to 25) he had won 25 titles. Their manager until very recently was Martin O'Neill, who was born in 1952. The words *Glasgow Celtic Football Club* contain 25 letters. Using the code $a = 1$, $b = 2$, $c = 3$, and so on, the sum of the letters in the word *Celtic* equals 52. Celtic's most glorious season was in 1967, when they won every competition that they entered. That year also they became the first British side to become champions of Europe, when they won the European Champions Cup on May 25. The sum of the digits in that famous date, 25/05/67 as you would write it, is 25."

"And the digits in 5/25/67, as you would write it, also sum to 25," I put in, to keep from drowning in the flood of 25s.

"Yes," said Richard, "another remarkable coincidence. That date was the 24251st day of the century: 24251 has 25 in its middle and its extremities sum to 25. $1888^{0.52} = 50.5267\ldots$. The reverse of the first four digits is 2505, May 25th again, and the next two digits give the year of the victory. That

Celtic team that brought the European Cup to Glasgow, incidentally, was an all-Scottish team consisting of players born within 32 miles of Glasgow. Of course, $32 = 2^5$."

"Richard, how can you possibly know so much about a Glasgow football club?" I asked.

"Read, Owen, read, and you will learn," Richard replied. "I came across a pamphlet on the history of the club. That and a pencil were all that is necessary. I must admit to having used a calculator to raise 1888 to the power 0.52. To be sure, I could have done it mentally with logarithms but as I age I see more clearly how finite life is and I take shortcuts. Not in the pamphlet but in the newspapers was the announcement of Martin O'Neill's resignation as Manager on 5/25 2005, exactly $25 + 2 \cdot 5 - 2 + 5$ years after the club was named champions of Europe.

"Had I been in Ireland then I would have bet on Celtic to win the European Champions Cup. The numbers were indicating that 1967 was going to be a special year for Celtic: $1967 = \dfrac{(2+5)!}{2.5} - 2 + 5 - 52$ and $67 = 6 \cdot 7 + 25$. That year Celtic also won the League, the Cup, and the League Cup. Yes, '67 was Jock Stein and Celtic's year. Did you know that if we use the code $a = 1, b = 2, c = 3$, and so on, *Stein* is 67? Did you know that the first two digits of $24251^{0.1888}$ are 67? Celtic was the number 1 team in Europe in '67—did you know that when 1888 is multiplied by its reversal 8881, the first three digits of the product are 167?"

"No, I didn't know any of those things," I said. "Who could? Until now, no one but you did. But tell me, are the numbers indicating any good bets now?"

"Ah, Owen, perhaps, perhaps," he said, glancing around cagily. He clearly was not going to give anything out. I took it no further. If the numbers had been giving him any good tips, he could have used the proceeds from them to improve his wardrobe. I changed the subject.

"You know," I said, "I have often thought that there is a great bond between the peoples of the U.S. and Ireland. I assume it goes way back to the 1840s, when America gave refuge to many Irish fleeing from hunger in Ireland."

"Indeed there is an affinity between our two countries," he said. "Consider: Ireland contains $32 = 4^2 + 4^2$ counties, and the U.S. contains $50 = 5^2 + 5^2$ states. The area of Ireland is 32,000 square miles and 32,000 acres is 50 square miles. The 31st and 32nd decimal digits of π are 5 and 0. The product of 32 and 50 is 1600, the address of the White House on Pennsylvania Avenue. The northern tip of Ireland lies very close to latitude $55.5°$ north, and $55.5 \cdot 32 = 1776$. The name of Ireland's capital, *Dublin*, begins with the fourth letter of the alphabet and *Washington* begins with the fourth letter from the end of the alphabet. Using the usual $a = 1, b = 2, \ldots$ code, the sum of the letters in the abbreviation *Ire* (for Ireland) is 32, the number of counties in Ireland, and the sum of the letters in *America* is 50, the number of states in the U.S.

"The number five," Richard went on, "seems to be a common link between Ireland and the U.S. The digits of 32 sum to 5, and 32 is 2 raised to power 5. There were originally 5 provinces in Ireland. Our Declaration of Independence was drafted by 5 men. The number of states originally in the U.S. was 13, or 2 raised to the third power, plus 5. The number of states now is $2 \cdot 5 \cdot 5$. There are just 5 denominational coins in circulation in the U.S. with less value than a dollar, and the motto *In God We Trust* was added to them as a result of legislation passed in 1955. The 50 states cover 5 time zones. There is just one time zone in Ireland, but it is 5 hours ahead of Eastern Standard Time. The Gaelic name for Ireland is *Eire*, whose first and last letter is the 5th letter of the alphabet. The Irish Constitution was adopted in 1937, and $37 = 2^5 + 5$. The letters in *USA* are the 21st, 19th, and 1st letters of the alphabet, the sum $21 + 19 + 1 = 41$ and $4 + 1 = 5$. In 1976 the U.S. celebrated its bicentennial, when it was $5^3 + 5^3 - 5^2 - 5^2$ years old. That same year 55 years had passed since the British-Irish Treaty offered independence to part of the island of Ireland. When the U.S. obtained its independence in 1776 there were initially 13, or half of $5^2 + 1$ colonies. Eventually, 37 or $2^5 + 5$ more states were added to the Union. Irish independence for 26, or $5^2 + 1$, counties was obtained in the Treaty in 1921. Northern Ireland consists of 6 or $5 + 1$ counties. Eire obtained its independence 145 or $5^3 + 5^2 - 5$ years after the U.S. achieved its independence. There are 5 Great Lakes in the U.S. The U.S. military headquarters is the Pentagon, a 5-sided building that consists of 5 layers of buildings on each of its 5 sides, each having 5 stories. Its 5-sided center courtyard covers 5 acres. There are 5 branches to the U.S. military: the Army, Navy, Air Force, Marines, and Coast Guard. There are 5^2 windows in the crown of the Statue of Liberty. Ireland is the 20th ($5^2 - 5$) largest island in the world. The population of the thirty-two counties of Ireland in 2004 was about 5 million. Every year in New York the Saint Patrick's Day Parade goes down Fifth Avenue."

"And everyone in both Ireland and the U.S. has 5 fingers on each hand," I put in, to stop temporarily the flood of 5s. But when Richard is in full spate he cannot easily be dammed up.

"My friend, 5 is an unusual number. There are 5 Platonic solids, and one of them, the dodecahedron, has 5-sided faces. The smallest integer that is both leg and hypotenuse of a primitive Pythagorean triangle is 5, in the (3, 4, 5) and (5, 12, 13) triangles. The 5th Mersenne prime, $2^{13} - 1 = -1 + 2^3 \cdot 4^5$, a representation with the first 5 digits in order. The 5th digit of π is 5. The 5th

Fibonacci number is 5. The 5th prime and the 5th Lucas number are both equal to 55/5. 5 is the sum of the primes that are less than it, and of the non-primes that are less than it. The number of seconds in one hour divided by 5! is 5 primorial. (In analogy with n factorial, the product of all the integers from n down to 1, n primorial is the product of all the primes from n down to 1, so 5 primorial is $5 \cdot 3 \cdot 2$. The word doesn't appear in dictionaries yet, though it's a good one. Dictionary-makers don't read enough mathematics.) The 5th day of the 5th month in a non-leap year is the 125th day of the year, and $125 = 5 \cdot 5 \cdot 5$. Also, $5! + 5 = 5^3$. Humans have 5 senses. The maximum number of solar eclipses in any one year is 5. A U.S. 5-cent piece weighs 5 grams. There are 5 vowels in the English language. That may be enough about 5 for now."

"Yes, it probably is," I said. "What else do you have?" I could tell that he was only just getting warmed up.

"Ireland is surrounded by water, which brings a curiosity to mind. The geographical center of the 32 counties of Ireland, which contains 4 provinces, is at latitude very close to $53°$ north. $53 \cdot 4 = 212$, and there we have both the freezing and boiling points of water. Here's another: the U.S. is 1 nation, it originally consisted of 13 states, 37 were added over time, and the number of states at present equals $55 - 5$. Put those numbers together, 11337555, and notice that $11 \cdot 3 \cdot 37 + 555 = 1776$, the year of the Declaration of Independence. Or take Ireland: it consists of 1 island, there are 6 counties in Northern Ireland, 26 in Southern Ireland, the Irish flag contains 3 colors, green, white and orange, *Ireland* contains 7 letters, and its abbreviated form (Ire) has 3. Put those together to get 1626373 and then, obviously, $1 \cdot 6 \cdot 263 + 7^3 = 1921$, the year of Irish independence. Southern Ireland officially adopted its Irish name, Eire, in the 37th year of the 20th century and the letters of *Eire*, in the usual code, sum to 37. Speaking of independence, the date of the American Declaration of Independence was 7/4 1776 and the first two digits of ln(1776) are 7 and 4. Also, the first two digits of ln (741776) are 1 and 3, for the 13 colonies."

"Are there any numbers that are *not* connected?" I wondered.

"Humph," Richard replied. "You think that finding these things is easy? You've heard violinists produce lovely sounds with their bows but if you were to try, all that you could evoke would be excruciating squawks. Those with no talent too often assume that artists produce their effects with no effort. Let me tell you emphatically, Owen, that that is not the case. Throughout the ages, creative people have been ignored by the philistine masses, scoffed at, given no credit—"

"Yes, yes," I said, before self-pity could get out of hand. "It was an idle comment, meant to be admiring. What next?"

"Here are two square numbers, with four and seven consecutive 5s in them respectively: $2357^2 = 5555449, 745356^2 = 555555566736$."

"Do you think that there are squares that have even more 5s?" I asked.

"Certainly there are," Richard said. "Even a non-artist could find them. Take the square root of, say, 55555555000000000000, round up to the next integer, square it, and you'll have a square starting with eight 5s. The artistry comes in finding those with as few non-fives as possible.

"Five 5s twice is 50, and 50 is the sum of five primes: $2 + 5 + 7 + 17 + 19$. That brings to mind a little alphametic puzzle that you may wish to give to your readers. Each of the letters represents a distinct digit in the following addition sum."

The professor wrote

$$
\begin{array}{r}
\text{FIFTY} \\
\text{STATES} \\
\hline
\text{AMERICA}
\end{array}
$$

"Well, A will have to be 1, so S will be 9, that means Y has to be 2, and—"

"Not now, Owen. Let your readers do it. The solution, I'm glad to say, is unique. Some composers of alphametics are actually letting some with more than one solution be published. They should have more taste. When Henry Dudeney made up the first one,

$$
\begin{array}{r}
\text{SEND} \\
\text{MORE} \\
\hline
\text{MONEY,}
\end{array}
$$

he didn't allow it to have extraneous solutions.

"Incidentally Owen, where were you at precisely fourteen thousand, six hundred and forty four seconds past midnight on Sunday, April 4, 2004?"

"I have no idea. I would guess that I was in bed sleeping. Why do you ask?"

"If you were asleep," said Richard, "you missed something special. Palm Sunday in 2004 fell on April 4. The time I just mentioned is unusual because then it was precisely 4 seconds past 4 minutes past 4 a.m. on the 4th day of the 4th month in the 4th year of the century. In addition to all of that, Palm Sunday falls on the 40th day of Lent. (The forty-six day period known as Lent is from Ash Wednesday to Easter Eve, both dates inclusive, but Sundays are not counted.) That won't happen again until the year 3204.

"I have to leave shortly, Owen but before I go I will give you some nice number patterns involving sums of triangular numbers, 1, 3, 6, ... The nth triangular number is $T_n = \dfrac{n(n+1)}{2}$. It is a fact that

$$T_1 + T_2 + T_3 = T_4,$$
$$T_5 + T_6 + T_7 + T_8 = T_9 + T_{10},$$
$$T_{11} + T_{12} + T_{13} + T_{14} + T_{15} = T_{16} + T_{17} + T_{18},$$

as can be checked by addition—the first is $1 + 3 + 6 = 10$. Are those equalities not pleasing? The pattern continues. It might be too much to ask of your readers to prove that, but who knows? Talent can spring up anywhere. Anyone who could take care of that would probably have no trouble showing that the pattern

$$2^4 = T_1 + T_5, \quad 3^4 = T_5 + T_{11}, \quad 4^4 = T_{11} + T_{19},$$
$$5^4 = T_{19} + T_{29}, \quad 6^4 = T_{29} + T_{41},$$

also continues. (The first subscript in the nth equation is $n^2 + n - 1$.) Here is one last pattern:

$$3(1+3)^3 = 1^3 + 3^3 + 2(1^4 + 3^4),$$
$$3(1+3+6)^3 = 1^3 + 3^3 + 6^3 + 2(1^4 + 3^4 + 6^4),$$
$$3(1+3+6+10)^3 = 1^3 + 3^3 + 6^3 + 10^3 + 2(1^4 + 3^4 + 6^4 + 10^4),$$

and so on and on, world without end."

"Those are very nice, Richard," I said.

"'Very nice' is faint praise, Owen." Richard said. "They are more than that. They illustrate the overwhelming elegance and harmony that exists in mathematics. Mathematics, the most glorious pursuit of the human intellect! Were I given to hyperbole I would say that doing mathematics is the *purpose* of the human race, but I am not given to hyperbole so I won't say that. Think, Owen, of what else must exist out there. What will we find next? But let us descend from the sublime. The human frame can stand only so much.

"Consider the 1992 Martell Grand National horse race in Liverpool. The winner was Party Politics, perhaps because the race occurred in the middle of a British general election campaign. The race was run on Saturday, April 4. April is the 4th month. The year 1992 is evenly divisible by 4. The race was scheduled to begin at 4 p.m. Its distance was 4 miles and 4 furlongs. There were 40 horses in the field but only 22 finished ($2 + 2 = 4$, as you have no doubt already observed). The winning horse was number 8 ($4 + 4$) in the line-up and went off at odds of 14 to 1. There are 4 divisors of 14. Last, but by no means least, each and every one of the horses running in the Grand National that day had 4 legs."

"Richard, you are crazy!" I said. "Saner people have been declared mad."
"Thank you for the compliment," said Richard.

Solutions

1. The professor's problem concerning zero temperature is solved as follows: A temperature twice as cold as zero has no meaning whatever. Zero multiplied by any number always equals zero. To say a temperature is twice zero is equivalent to saying that one has no money in one's pocket, but someone else has twice as much in his or her pocket.

2. Before solving the professor's second problem, recall that, if F and C denote Fahrenheit and Celsius degrees, then

$$F = \frac{9}{5}C + 32 \quad \text{and} \quad C = \frac{5}{9}(F - 32).$$

So, we want to find n so that

$$n = \frac{9}{5}n + 32 \quad \text{or} \quad n = \frac{5}{9}(n - 32).$$

To solve the second equation for n, we have

$$9n = 5(n - 32), \quad \text{or } 9n = 5n - 160, \quad \text{or } 4n = -160, \text{ so } n = -40.$$

Solving the first equation would give the same result. Thus, at $40°$ below zero, you are exactly as cold whichever scale of measurement you use.

3. The professor's third problem can be solved as follows: call the Fahrenheit temperature t, let its first digit be d, and suppose that it has n digits. Let us see what we do when we move the first digit of t to its end. For example, when we change 527 to 275, we are subtracting 500, multiplying 27 by 10, and adding 5 to get 275. In general, we are forming

$$(t - 10^{n-1}d) \cdot 10 + d.$$

So, if we want that to be same as the Celsius temperature corresponding to t, we are trying to find t, d, and n so that

$$(t - 10^{n-1}d) \cdot 10 + d = \frac{5}{9}(t - 32).$$

That is,

$$10t - 10^n d + d = \frac{5}{9}(t - 32).$$

Multiply by $\dfrac{9}{5}$:

$$18t - \frac{9}{5} \cdot 10^n d + \frac{9}{5}d = t - 32 .$$

Let us take $d = 5$. The equation becomes

$$17t = 9 \cdot 10^n - 41 .$$

So, $17t$ is $899\ldots959$, with $n - 2$ 9s in its middle. We know that t starts with a 5 ($5 \cdot 17 = 85$, and $6 \cdot 17 = 102$ is too big) and must end with something that, when multiplied by 17, has last two digits 59. Looking at a 17-times table, or writing one out if you don't have one handy, we see that the last two digits of t have to be 27. Thus we have discovered our first solution, 527. To discover the next, we have to fill in the blanks in the multiplication

$$
\begin{array}{r}
5\ldots\ldots\ldots\ldots 27 \\
17 \\
\hline
89\ldots\ldots\ldots 9959.
\end{array}
$$

We do this by discovering the unknown digits of t one by one, from right to left. Suppose that

$$
\begin{array}{r}
5\ldots\ldots\ldots\ldots r27 \\
17 \\
\hline
89\ldots\ldots\ldots 959.
\end{array}
$$

Then we have

$$
\begin{array}{r}
5\ldots\ldots\ldots\ldots r27 \\
17 \\
\hline
\ldots\ldots (7r + 1)89.
\end{array}
$$

So, what do we have in the third column from the end? A 9 at its bottom, so $7r + 1$, plus 2, plus the 1 that is carried from the second column from the end, has to be 9, possibly with something to carry. So, $7r + 4$ is 9 plus a multiple of 10 (the multiplier may be 0), $7r$ is 5 plus a multiple of 10, and this is possible only if $r = 5$. Now keep going and determine the digit immediately to the left. (Once you get into the swing of determining digits, it becomes quick.) Continue determining digits and eventually you will have a solution, which is where the professor's 52941176470588235 27 came from.

4. The unique solution to the alphametic puzzle is

$$
\begin{array}{r}
65682 \\
981849 \\
\hline
1047531.
\end{array}
$$

Reference for further reading

David Wells. *The Penguin Dictionary of Curious and Interesting Numbers.* Penguin Books, 1986.

CHAPTER **4**

Curiosities in armed conflicts

We were in Galway for the annual oyster festival. When Richard ordered—I mean suggested—that we go to it, I naturally asked why.

"Owen, you have an unfortunate tendency to be a stick-in-the-mud, and I see it as my duty to do all that I can to counteract that," Richard said. "Besides, I'll wager that you, a native of Ireland, have never even been to your justly renowned oyster festival in the colorful city of Galway. Am I right?"

I had to admit it.

"There you are, I win my bet," Richard said. "You, the loser, have to provide transportation and expenses. In return, your cultural horizons will be expanded and I may have some new items to give you for your column."

"Those will be worth having," I said, "even if we have to go to not-so-colorful Galway."

"You're right and wrong, Owen," the professor said. "The items will indeed be worth having, and in the late 1950s the Bishop of Galway was named Brown, the Mayor was named Green, and the Harbor Master was named Whyte."

There were quite a few people at the festival, most of them younger than I. I discovered that it had been in existence for more than fifty years, originally being the invention of a clever promoter seeking to extend the tourist season past August into a month with an *r* in it.

"Consider the oyster, Owen," Richard said, "and be glad you are not one. Those very few baby oysters who survive three weeks of floating freely in the sea feel the need to go to the bottom and attach themselves to something, never to move again. Their only entertainment is to change from male to female and back again, which they do at least once. They used to be so common that they made up a major part of the diet of the poor, but the species has fallen on hard times. What the privileged classes would once turn up their noses at they now pay through the nose for. You can't use that in your column, Owen, I'm saving

51

it for my monograph on the uses of the nose. Let's go see the oyster opening contest."

"What for?" I asked. It was crowded, and I was tired.

"Owen, sometimes I despair," Richard said. "These shuckers are the best in the world. The U.S. representatives have to qualify at the Nationals in Maryland. No Irishman has won since 1996. France, Switzerland, Sweden, Canada—that's where the winners come from now. Don't you like to see people doing something really well? Especially when it's done for its own sake. Oyster shuckers, even the very best, don't get money or fame, not even fifteen minutes worth. They're similar to recreational mathematicians that way."

We went, and saw people doing things to oysters very quickly.

"Thirty oysters, Owen, to be opened in as little time as possible, with penalty points added for errors. Shell or grit on the surface of the oyster—4 points. Blood—30 points."

"Whose blood?" I asked.

Richard paid no attention. "Last year the winning time, after penalties, was three minutes and one second. Think of the skill!"

"What shall we eat after the contest?" I asked. "Oysters?"

"Certainly not," Richard said. "We are above peasant food."

When we were dining (not on oysters) I said, "I was very pleased to obtain the 9/11 curiosities. Thank you so much. I was wondering if you had any curiosities concerning wars or conflicts."

"Certainly. Wars are always with us, as are numbers." Richard said. "Take the First World War. The year it began, 1914, is evenly divisible by 11. The event that triggered the First World War was the assassination of the heir to the Austro–Hungarian throne, the Archduke Franz Ferdinand. He and his wife were shot at 10:50 a.m. on the morning of Sunday, June 28, 1914. The Archduke's wife died almost immediately. The Archduke died ten minutes later, at the eleventh hour. And the war ended at the eleventh hour of the eleventh day of the eleventh month in 1918.

"Moving on to the Second, and let us hope the last, World War, it is generally agreed to have commenced on September 3, 1939, when Germany invaded Poland. The date 9/3/39 is palindromic. Not everyone has noticed that the Second World War lasted exactly six years—on Sunday, September 2, 1945 Japan signed the formal surrender document on the quarterdeck of the battleship USS Missouri in Tokyo Bay. The war lasted exactly 313 weeks, another palindrome. Partition 313 as 31/3 and as 3/13. $31 \cdot 3 = 93$ and 9/3 gives the month and day that the war started and $3 \cdot 13 = 39$, the year the war commenced.

"You know that Neville Chamberlain resigned as Prime Minister in May, 1940. He actually resigned from Number 10 Downing Street on May 10. That was the 131st day of that year (1940 was a leap year).

"Another palindrome," I said.

"All of this talk about the Second World War brings to mind two remarkable coincidences that occurred during that terrible conflict," Richard said. "The first concerns the invasion of Normandy on the northwestern coast of France by the Allied Forces on D-Day, which was Tuesday, June 6, 1944."

"I recall reading," I said, "that that invasion involved the largest deployment of armed forces and naval ships (up to that period) in history."

"That is correct," said Richard. "In the weeks and months prior to the invasion the Allies main aim was to keep secret the details of the planned invasion of northwest Europe."

"The Nazis expected any such invasion to come across the English Channel from Dover," I said. "The Allies were aware of that, of course, and obviously wanted the element of surprise. Therefore, details of any such invasion would have to be kept secret."

"Precisely," said Richard. "That is why the Allied leaders were shocked when it was brought to their attention that a series of crossword clues and answers in *The Daily Telegraph* shortly before D-Day contained top-secret codenames that were only known to the Allied leaders."

"Codes hidden in crossword clues and answers! You are not serious," I said.

"Oh, but I am," said Richard. "Two answers to the *Telegraph* crossword were *Utah* and *Omaha*. These were codenames for the beaches on which the Allied forces were planning to land on in Normandy. Later, another crossword answer was *Overlord*, which was the codename for the entire Allied invasion plan. The British intelligence service, MI5, were reportedly worried that the crossword column of the *Telegraph* was being used to convey secretly top secret information to the Nazis. They interviewed the crossword compiler, a fifty-four year old teacher from Leatherhead, in Surrey, named Leonard Dawe. He had been the Telegraph's crossword compiler for the previous twenty years, and was reportedly a quiet, unassuming man. Dawe said that he had no idea why he chose those particular words for his crossword solutions. He also maintained that he did not know that those words were codenames.

"Then there was the case of the two advertisements in *The New Yorker* on Saturday, 22 November 1941, promoting a new dice game called 'The Deadly Double'."

"What was unusual about that?" I asked.

"On page 32 of the magazine on that day," said Richard, "a boxed advertisement appeared. The advertisement was headed *Achtung Warning Alerte*! Underneath was written: *see advertisement, page* 86. Underneath this caption was a picture of two dice, one white and one black. On the white die were the numbers 12 and 24, and the letters XX. On the black die were the digits 0, 5 and 7. These dice numbers and letters have been interpreted as bearing the following secret message: 0 hour for a double cross on the 7th day of the 12th month at the 5th hour out of 24. The Japanese attack on Pearl Harbor occurred on the 7th of December, 1941, at approximately 7.00 a.m. (It may have been originally intended for the attack to occur at 5.00 a.m.) The year of the attack may also have been hidden in the numbers on the two dice. Let the numbers on the white die, 1224, represent a year. Add to this the sum of the digits on the black die: $1224 + 12 = 1236$. Then add to this 705, which appears on the black die. The result: 1941, which is the year of the attack on Pearl Harbor.

"The second advertisement on page 86 shows a mixed group of people playing the dice game during an air raid, while outside searchlights are directed towards the night sky as an explosion occurs on the ground. At the bottom of the advertisement was a symbol similar to that of the German double-headed eagle. The letters XX (20 in Roman numerals) are repeated on this symbol. The approximate latitude of Pearl Harbor is 20 degrees north. In 1967 it was revealed that the FBI had investigated the matter in 1941, but had concluded that the whole episode could be put down to one big coincidence."

"Yes, you'd wonder why anyone would go to such trouble, and for what reason," I said. "Have you any coincidences concerning other armed conflicts?"

"Indeed I have," said Richard. "Consider the Irish conflict. In 1922, the leader of the Irish Free State Army was Michael Collins. Ireland was engulfed in a terrible civil war at the time. Collins was in a classic catch-22 situation. Is it not ironic that Collins was shot on August 22, in the year of 22? Putting the two 22s together gives 2222. 2222 in base 8 is 1170, the year Ireland was first invaded by the British. Incidentally, 2222 is an interesting number in its own right, because $2222 = 2^1 + 2^2 + 2^3 + 2^5 + 2^7 + 2^{11}$ and the total number of 2s in that equation is 11, the largest of the prime exponents. Excluding the exponent 1, the five other exponents are the first five primes in order.

"Michael Collins was 31 years old at the time of his death in 1922. The four digits, 2, 2, 3 and 1 can be arranged to give $2 \cdot 31^2 = 1922$. Collins was Commander-in-Chief of the Irish Free State Army. Among the military leaders, he was number 1 in Ireland. Ireland's number 1 was shot on the 234th day of the year."

"The first four digits in order," I said,

"Yes," said the professor. "And $34 - 12 = 22$. Add the fifth digit: $1 \cdot 234 \cdot 5 = 1170$, that ill-starred year when the British invaded Ireland. If we use the nine digits in ascending order, and the plus, minus, and multiplication signs of arithmetic only, there are just two ways to get 1922:

$$12 + 3 + 45 \cdot 6 \cdot 7 + 8 + 9 = 1922$$
$$12 + 34 \cdot 56 + 7 + 8 - 9 = 1922.$$

If we allow a minus sign in front of the first digit, and using the plus, minus and multiplication signs, one finds that there is a unique expression for 1922. Your readers might enjoy searching for it."

"I am sure they will," I said.

"Collins was killed," said the professor, "in a place in County Cork known as Beal na Blath (if you put it in your column, you had better say that it is pronounced bale-naw-blaw), which is Gaelic for Valley of the Flowers. The second letter of the alphabet is B, so the two B's in Beal na Blath represent the two 2s in 22, the year of the assassination. Incidentally, 22 is the sum of two different arithmetic progressions: $1 + 4 + 7 + 10$ and $4 + 5 + 6 + 7$."

"And $22 = -2 + 3 + 8 + 13$, another arithmetic progression," I said.

"Yes, and it's $-11 + 0 + 11 + 22$, and $3 + 4\frac{2}{3} + 6\frac{1}{3} + 8$ as well, but we must have some standards," the professor said. "Not all arithmetic progressions are equally good, you know. When Collins was shot the news of the ambush and killing went around Ireland at lightning speed. The ordinary people of the country—gripped as they were in the middle of a bitter civil war—were distraught at the death of Ireland's military leader. Their heartfelt plea the next day is summed up by the following:

Ambush. Beal. Collins dead. End fighting. God help Ireland."

"The first nine letters of the alphabet, in order," I said.

"Correct," said Richard. "Collins fought in the Easter Rebellion in Dublin in 1916. The last headquarters of the rebel leaders before surrendering to British forces was at a house in Moore Street, in Dublin. That house was numbered 16. The leader of the rebellion was Padraig Pearse. Note that the initials of both his first name and surname begin with the 16th letter of the alphabet. Here's 1916 using the digits in descending order: $-9 - 8 - 7 + 6^5/4 - 3 - 2 + 1 = 1916$. Ask your readers if they can use the digits 1 through 9 in ascending order, inserting plus and minus signs wherever they wish, to get 1916. There is only one solution. There is no solution with a minus sign in front of the first digit, but with a minus sign in front of the first digit, and using addition, subtraction, and multiplication, there is just one way to represent 1916. Ask your readers if they can find it.

"You may find it interesting to note that the first two digits of the 16th root of 1916 are 1 and 6. Also, 1916 is the sum of six consecutive primes: 307, 311, 313, 317, 331 and 337. The 1916 Rising commenced on Easter Monday, April 24. The Rising was planned for Easter time for symbolic reasons. The Rising, and the rebirth of the Irish nation it claimed it hoped to foster, could be compared with the Resurrection and with the rebirth of nature in spring. But the date of the Rising, April 24, is curious. The geographical center of Ireland is very close to $8°$ west and $53°$ north. $8 \cdot 53 = 424$—April 24th. The Irish War of Independence is generally assumed to have officially commenced on the 21st day of January, 1919. That date, 21, is the $(1 + 9)$th prime, plus $1 - 9$. You may also be interested to know that one of the most famous encounters between the British and Irish forces occurred at Cross*barry* in County Cork, in 1921. The leader of the Irish forces that day, March 19, 1921, was General Tom *Barry*. The date may be written in the European style as 19/3/21 or 19321. The battle played a central role in the Irish War of Independence. Using the usual code $a = 1$, $b = 2$, $c = 3$, and so on, the sum of the letters in *Crossbarry* is 138. Since it is the number 1 battle in the War, add 1 to get 139. The central digit, 3, gives the month of the battle and the outside digits 19 give the day. The square of 139 is—you figure it out, Owen."

I did. "It's 19321, the date of the battle," I said.

"Uncanny, is it not?" the professor asked. "Two major political events took place in Ireland in the nineteenth and twentieth centuries, the Act of Union between Britain and Ireland in 1801, and The British–Irish Treaty in 1921. Note that $1801 = 24^2 + 35^2$ and $1921 = 25^2 + 36^2$. The numbers that are squared go up by 1."

"So we should look for something momentous in $26^2 + 37^2$," I said. "Let's see, that's 2045. What could it be?"

"Waiting for it will give you something to live for, Owen," Richard said. "But you missed an intermediate event, the announcement of the IRA ceasefire in 1994. 1994 is $25^2 + 37^2$. Four years later, in 1998, the major political event in Ireland was the signing of The Good Friday Agreement, and 1998 is $1^3 + 9^3 + 9^3 + 8^3 + 1 + 9 + 9 + 8$."

"Richard, how can you know so much about Ireland?" I asked. "You don't live here."

"The active mind ranges widely," Richard said, barely keeping superciliousness in check. "I am a long-time Irishophile, and in addition I prepared for this visit. The prepared mind appreciates more deeply, don't you think?"

"Certainly," I said. I should never have asked. "But go on, if you have more."

"Of course I do. Ireland is not quite as inexhaustible as mathematics, but there is enough for at least several lifetimes. The first British soldier to be killed in Northern Ireland since the most recent troubles began was Gunner Robert Curtis, who came from Newcastle on Tyne, in the north of England. A man named Billy Reid, who was a member of the Irish Republican Army (IRA), killed him in February 1971. Three months later Billy Reid was shot dead by the British Army in a street in Belfast—Curtis Street. An amazing coincidence occurred when the IRA announced their ceasefire in August 1994. As you know, the first ceasefire broke down, and the military conflict resumed, before a second ceasefire was called."

"Yes," I said.

"Well," said the professor, "the first ceasefire commenced at midnight on Wednesday, August 31, 1994. It ended at 1800 hours on Friday, February 9, 1996. The second ceasefire commenced at 1200 hours on Sunday, July 20, 1997. The first ceasefire lasted for 526 days and 18 hours exactly. The period between the two ceasefires also lasted for 526 days and 18 hours exactly."

"That's incredible," I said.

"On the contrary," the professor said. "It occurred, so it is completely credible. After an event happens, you no longer have the privilege of disbelieving in it. You're a writer, Owen, you really should use words with more care."

"*Irishophile* is a barbarism, you know. You really shouldn't mix English and Greek," I said. It was the best I could do.

"Correct English is that which is used by educated speakers of the language. I am an educated speaker of the language. Therefore what I speak is correct English," Richard said. I thought that there was a logical flaw there, but I could see that he was determined to have the last word, so I didn't pursue it.

"As we approach Easter 2005, what do you think of the present situation in Ireland, Richard?" I asked. "It is said that there is a good working relationship between the Irish and British Prime Ministers. Surely that is a good omen for the future."

"Yes, it is," said the professor, "Of course, such a good relationship is to be expected. The Irish Prime Minister is Bertie Ahern, and though his British counterpart likes to be known as Tony Blair, he is actually Anthony Blair. Initials B. A. and A. B. are mirror images of each other, and so a good signal that they will see eye to eye."

"Incidentally," I said, "I recall you mentioning some numerical curiosities concerning the 9/11 attacks and the Bali bombing in 2002. Have you any other curiosities concerning the Bali bombing?"

"Indeed I do, Owen. First of all, let me point out that 202 people were killed in the Bali bombing, which occurred on October 12, 2002. That date can be written as 10/12/02, or as 101202. The number formed by the first three digits is half the number formed by the last three digits, and those are the number killed in the attack. Notice that $202 = \frac{20^2 + 2^{02}}{2}$. Pretty, isn't it? You can't do that with many integers. Also, $2 + 0 + 2 = 4$, the first four primes are 2, 3, 5, and 7, and $(2 + 3 + 5 + 7)^2 - (2^2 + 3^2 + 5^2 + 7^2) = 202$. There's more. Write the dates of the two attacks as 91101and 101202. The difference between the two numbers is 10101, which a palindrome. The Bali attack occurred in the year 2002, which also is a palindrome. The number killed in that attack is a palindrome. Those two atrocities may be referred to as the 9/11 attacks and the 10/12 attacks. These four numbers, 9, 10, 11 and 12, are consecutive. The last three digits of the following sum, $9^4 + 11^4 = 21202$, give the number killed in the Bali attack. The natural logarithm of 10122002 (the date of the Bali attack written in full) is 16.130222028. . . and there is a 202 within that decimal expansion."

"But the digits 202 will appear in the logarithm of almost anything," I said.

"Indeed they will," the professor said, "but not that soon. On the average, only one in a thousand three-digit sequence will be 202, so you wouldn't expect to see one until around the 500th place of the expansion. Something must have been pushing 202 to the front. Further, $(\frac{202}{2})^{.202} = 2.54022 \ldots . 254$ is the day number of the year in which the 9/11 attacks occurred."

"Do you have any curiosities on the U.S.–Iraq war?" I asked.

"I certainly do, Owen, but far too many to mention now. I'll tell you later. In any event, it's time to hit the road. An odd expression, that. 'Join the road' would be more accurate, but there's no accounting for language. I have a route planned that will take us through some parts of the country that I want to see. You may even enjoy it—many of the roads are paved."

We left, taking no oysters with us.

Solutions

1. The reader was asked to place plus, minus, and multiplication signs in the ascending sequence 1 2 3 4 5 6 7 8 9, and to also place a minus sign in front of the first digit, to get 1922. There is only one solution: $-1 + 234 + 5 \cdot 6 \cdot 7 \cdot 8 + 9 = 1922$

2. The reader was asked to insert the plus and minus signs anywhere in the ascending sequence 1 2 3 4 5 6 7 8 9, so that the expression equals 1916. There is a unique solution, $1234 - 5 + 678 + 9 = 1916$. With a minus sign

in front of the first digit, there is also only one solution: $-12 + 34 \cdot 56 + 7 + 8 + 9 = 1916$.

References for further reading

Martin Gardner, *Knotted Doughnuts and Other Mathematical Entertainments.* W. H. Freeman, 1986, Chapter 1, "Coincidences."

Ken Anderson, *Coincidences. Chance or Fate.* Blandford, 1995, pp. 62–63.

Martin Plimmer and Brian King, *Beyond Coincidence.* Icon Books, 2004.

CHAPTER 5

Number and word palindromes

"Owen, you have three days to spare for a trip, don't you?" Richard asked.

"Well, not really," I said.

"Of course you do! You've probably never been to Skreen, have you?"

"No," I said, "Why would I want to go there?"

"Owen, your provincial attitude will be the death of me. There is more to life than Cobh and Cork, you know. The world, even Ireland, is large. You should get about in it while you're still able. Skreen is the site of the Stokes Summer School, and I want to attend one of the lectures."

"You have to go to summer school because you failed something during the year?" I asked.

"I acknowledge your attempt at humor, Owen." Richard said. "No doubt it is the best you can do. The Stokes Summer School is an annual gathering in honor of one of Ireland's greatest mathematicians, George Stokes, who was born in Skreen in 1819."

"But he went to England, didn't he?" I said.

"Very good!" said the professor. "He did indeed. The major export of Ireland in the nineteenth century was people. Many people of ability had to leave in order to use their talents properly—quite a few of them leaving from Cobh— and Stokes was one of them, though it's likely that he departed from Dublin. Off he went to Pembroke College in Cambridge, excelled in mathematics and was elected to a fellowship when he graduated. Schools don't hire their graduates any more. The system worked very well for Oxford and Cambridge back then, but I suppose that times have changed."

"Though not necessarily for the better," I said, knowing that Richard would like to hear it.

"Too true," he said. "Pembroke threw him out in 1857 because he got married, but in 1869 the rules changed and he was let back in. He was elected Master of the College the day before his 84th birthday. Scholars were not turned out to pasture then as they are now. He had a long and distinguished career as a mathematical physicist, and he has achieved immortality by having Stokes's Theorem named after him."

"What's Stokes's Theorem?" I asked.

"It's much too technical for your readers, Owen," the professor said. "Let's leave Thursday. Michelle will be coming too. While I'm at the school sessions, you can take her sightseeing. There's the Black Monument to see. I hear it has a sign near it saying 'No shooting tourists' so you'll be perfectly safe. I've been saving up items for your column, you know."

There was further discussion, but we ended up in Skreen. Our second evening there, we settled in, I with pencil and paper.

Richard showed me a magic square

$$
\begin{array}{ccc}
282 & 737 & 646 \\
919 & 555 & 191 \\
464 & 373 & 828
\end{array}
$$

made up entirely of palindromes.

"The central number in the bottom row," he said, "is an interesting palindrome. It's a prime, a sum of five consecutive primes, $67 + 71 + 73 + 79 + 83$, and a sum of the squares of five consecutive primes, $3^2 + 5^2 + 7^2 + 11^2 + 13^2$. And water boils at $373°$ Kelvin."

"Aha!" I said, "I see how the square was constructed. You took the single-digit square

$$
\begin{array}{ccc}
8 & 3 & 4 \\
1 & 5 & 9 \\
6 & 7 & 2
\end{array}
$$

and put around the digits a copy of

$$
\begin{array}{ccc}
2 & 7 & 6 \\
9 & 5 & 1 \\
4 & 3 & 8
\end{array}
$$

So of course you get a magic square. Could you repeat the process to get a five-digit square? And would its entries have any properties?"

"I'll leave the possibility of the construction to you and your readers" he said. "If it's possible, of course its entries will have properties. Some might even be interesting."

"Yes," I said, "that was what I meant."

"Then that is what you should have said," Richard said. "Words should be treated with respect. A *hippodrome* is a place where horses run, a *syndrome* is several symptoms running together, a *dromedary* is a running animal, so therefore a *palindrome* is?"

"I suppose something that runs backwards," I replied, my mind luckily still alert.

"Quite right," Richard said. "Here is another palindrome, 134,757,431, with what you would call a property: it is simultaneously

$$1^7 + 2^3 + 3^8 + 4^5 + 5^4 + 6^2 + 7^1 + 8^9 + 9^6,$$
$$1^7 + 2^5 + 3^8 + 4^1 + 5^2 + 6^4 + 7^3 + 8^9 + 9^6, \text{ and}$$
$$1^7 + 2^8 + 3^4 + 4^2 + 5^3 + 6^5 + 7^1 + 8^9 + 9^6.$$

Each of the digits appears exactly once as an exponent and as a base."

"Ah," I said, my mind still firing on all cylinders. "There are 9! such numbers, and that is, let me see . . . " I started to get out my calculator, which I bring to every meeting with Richard.

"362,880," Richard said immediately. "Haven't I told you, Owen, to make friends with numbers and they will make friends with you?"

"You have," I said, "but you have more friends than I. I wonder what the chance is that an integer would have three such representations, or more."

"That," Richard said, "is a topic for investigation. If you allow 0 among the digits, then there are 10! numbers to browse among. I won't ask you to calculate that."

Richard can be insufferable at times.

"The smallest palindrome containing all ten digits is 1023456789876543201, obviously. The smallest palindromic *prime* containing all ten digits," said Richard, "is

$$1023456987896543201.$$

Those of your readers skilled in computer programming might enjoy finding what the next smallest is, and the largest. The sum of the first 666 palindromic primes is 2,391,951,273, unfortunately not a palindrome, but

$$2^3 + 3^3 + 9^3 + 1^3 + 9^3 + 5^3 + 1^3 + 2^3 + 7^3 + 3^3 = 666 + 666 + 666.$$

The product of the five consecutive primes 7, 11, 13, 17, and 19 is 323,323, a palindrome and a nine-digit power sum, $1^7 + 2^2 + 3^8 + 4^9 + 5^5 + 6^6 + 7^1 + 8^4 + 9^3$. Since I see that you have your calculator out, you can tell me what the sum of the squares of the first nine consecutive odd integers is."

"It's 969," I answered, fairly quickly, "a palindrome."

"Just so," Richard said. Here are some more curiosities.

$$1234321 = 11^2 \cdot 101^2,$$

the product of two prime palindromes. Look at this product:

$$123{,}456{,}789 \cdot 989{,}010{,}989 = 122{,}100{,}120{,}987{,}654{,}321.$$

On the left 123456789, on the right 987654321, and the second factor on the left is a palindrome. On the right, 122100120 is unfortunately not palindromic, but does have the virtue of containing exactly three occurrences of the first three non-negative integers. Moving on, the palindrome 64446 is the sum of two consecutive primes, 32213 and 32233 and it is also the sum of their reverses, $31223 + 33223$."

"Just a moment," I put in. I really was exceptionally alert that day. "That's not so surprising. When you reverse things you often get the same sum. Using your number, $64446 = 31123 + 33323 = 32333 + 32113$ and I could make up many more examples."

"But with primes, and consecutive ones at that?" Richard said. "I think not. And I doubt that you could find another reversal like

$$203313 \cdot 657624 = 426756 \cdot 313302."$$

"You are right there," I had to admit.

"As I think I've mentioned before, the only palindromic prime with an even number of digits is 11," Richard said. No other palindromic number with an even number of digits can be prime."

"Yes," I put in quickly, "Because they're all divisible by 11."

"Very good, Owen," Richard said. "I wonder if your readers would be able to show that that is the case. While we are on the subject of 1s, it's interesting to note that

$$1111 = 11^2 + 12^2 + 13^2 + 14^2 + 15^2 + 16^2$$

and

$$11111111 = 1234567 \cdot 9 + 8.$$

I expect that you never noticed that

$$\frac{2 \cdot 22 \cdot 222 \cdot 2222}{2 \cdot 2 \cdot 2 \cdot 2} = 1356531 = 11 \cdot 123321,$$

the sum of the digits of 1356531 is $2 + 22$, and $123321 = 111 \cdot 1111$."

"Well," I said, exceptionally daring, "that last equality is no surprise, since it's just a rewriting of the original."

"You are *quick*, Owen," he said, a little annoyed at being caught out, "Perhaps even quicker than your readers. Here are two more palindromic facts, the last that I have: the palindromic prime 383 is a sum of three consecutive palindromic primes, $101 + 131 + 151$, and the smallest palindrome that is the product of four consecutive primes is $5005 = 5 \cdot 7 \cdot 11 \cdot 13$."

"Thank you, Richard," I said. "You've given me plenty of good material. What about letter palindromes? Like the one about Theodore Roosevelt, 'A man, a plan, a canal—Panama.' Do you know of any interesting ones?"

"Yes, indeed. By the way, the Panama palindrome can be altered, though not flatteringly to Roosevelt, to 'A man, a pain, a mania—Panama.' And other people have taken it to extremes like 'A man, a plan, a cat, a ham, a yak, a yam, a hat, a canal—Panama" and longer ones even more ridiculous. Someone has constructed (not by hand!) a palindrome with more than 17000 words. Mathematicians would not be so tasteless. When Peter Hilton, the well-known topologist, was challenged to construct a palindrome, 'Sex at noon taxes' was the result. It shows that talent in one field carries over into others. Brevity is the soul of good palindromes. 'And E. T. saw waste DNA' expresses one possible reaction of aliens to us. And if Neil Armstrong had seen E. T. while on the moon, why then, 'Neil A. sees alien.' "

"But Richard," I said daringly, "People say that puns on names are the lowest form of wit (as buns could be said to be the lowest form of wheat). Doesn't that carry over to palindromes?"

"What, you'd sacrifice the first words spoken in the Garden of Eden, 'Madam, I'm Adam,' and the demure reply, 'Eve'? And that in eating the apple, 'Eve damned Eden, mad Eve.'?"

"I suppose not," I allowed. "Palindromes with proper names are all right if they're *good* palindromes."

"Yes," Richard answered. "For example, an inhabitant of the island of Nauru is a Nauruan. It's possible, though perhaps not likely, that he or she speaks

Malayalam. Rubidium bromide, RbBr, should be used in mirrors. *Radar* is *macam* in Hebrew, a two-language palindrome. No three-language palindrome has been discovered yet as far as I know. The longest palindromic word in English is *tattarrattat*, though it probably appears nowhere else than in Joyce's *Ulysses*. But letter palindromes really aren't mathematical."

"That's true," I said. "Have you got anything else that might be of interest?"

"What a question, Owen! Of course I have more of interest. I *always* have more of interest. Sometimes you lack the time or the energy to absorb it. Here are some items involving playing cards. I've heard an old Irish saying that 'the devil is in the playing cards,' though I'm not sure why—it may have something to do with playing cards and drinking—and the Puritans thought that they were devilish items. I think that even now in the United States there are sects that forbid their members to touch them. After all, the sum of the letter values in *devil*, with $a = 1$, $b = 2, \ldots$, is 52. The sum of the letter values in *God* is 26, though what the import of that is isn't clear. No one could say that God is playing with half a deck. In any event, in

1	deck of cards there are
2	different colors,
26	cards of each color,
13	cards in each suit,
12	face cards, and a total of
52	cards.

These six expressions, all using those ten digits in the same order,

$$1 + 2 + 26 + 13 + 12 \cdot 52,$$
$$1226/(-1 + 3) - 1 + 2 + 52,$$
$$1 \cdot 2 \cdot 26 \cdot 13 - 1 - 2 - 5 - 2,$$
$$-1 + 226 \cdot 1 \cdot 3 + 1 - 2 \cdot 5 - 2,$$
$$1 \cdot 2 + 2 + 613 - 1 + 25 \cdot 2, \text{ and}$$
$$1 \cdot 22 + 613 + 1 + 2^5 - 2$$

all have the same value. Can you see what it is without your calculator, Owen?"

"The last two are both 666," I said. "I'll take your word for it that the others are too."

"Indeed they are," Richard said. "What's more, the sum of the integers from 1 to 52 is 1378. If we cut the deck (as it were) into 37 and 18, we have two integers whose product is . . . "

"Don't tell me," I interrupted, "it must be 666."

"As it is," Richard replied, with a satisfied smirk. "Did you know," he went on, "that Harry Houdini, the magician who did card tricks as well as made escapes from locked boxes and the like, was born on March 24, 1874 and died on October 31, 1926?"

"No, I didn't," I answered. "Now are you going to tell me that those are palindromes whose sum is 666?"

"Don't be ridiculous, Owen," he said. "Not *everything* can be marvelous. But Houdini did cut his deck exactly down the middle, living 26 years in each of two centuries, and he maintained his mystery by dying on Halloween, a day associated with mystery. Talking about cards brings dice to mind. The sum of the spots, $1 + 2 + 3 + 4 + 5 + 6 = 21$ and when you turn *dice* into a number, using the code $a = 1, b = 2$, and so on, the total is 21. By the way, there's a gambling game with dice that you should tell your readers to avoid. It may have died out by now, but I can remember playing it (illegally) at traveling carnivals when I was a child and knew no better. I wouldn't be surprised if it still exists in some of the casinos of the lower class. Players can bet on any of the numbers from 1 to 6. The swindler—the operator of the game, I mean—tosses three dice. When I played the game, the dice were oversized and in an hour-glass shaped cage, with a crank to rotate it. If all three dice show the same face, say three 6s, those who bet on 6 win three times their bet, and everyone else loses. If the dice came up with two 6s and a 3, those who backed 6 win twice their bet, and those who bet on 3 win the amount of their bet. If all three dice show different numbers, those who bet on those numbers win the amount of their bet."

"I suppose that players think that since each face has a one in six chance of coming up, when three dice are tossed there are therefore three chances in six that the number they bet on will appear," I said.

"Exactly, Owen! You are exceptionally acute today," Richard said. (I thought so too.) "And then the mark—the player, I mean—thinks that if his number comes up more than once, he wins even more, so that this is a very good game to play."

"But it isn't," I said.

"No, it is an exceptionally bad game to play. Your readers might enjoy working out the true odds. After they do they may stick to craps, where they can give away their money more slowly. Or, better yet, they can not gamble at all, unless they view it as entertainment and the amount that they lose as the price they have to pay for their fun. I can think of more rewarding forms of entertainment. Why, with modern slot machines you don't even get the benefit of the arm exercise you used to when you had to pull the lever over and over. You're more in danger of getting carpal tunnel syndrome from pushing buttons repeatedly. But I'm getting off the subject, and we've occupied this table quite long enough."

"As always, thank you, Richard," I said. "Even if you won't tell me what Stokes's Theorem is, you've given me enough to stoke my column's furnace."

"Quite so," said Richard, rolling his eyes upwards. "As always, I'll leave the check to you."

Solutions

1. One way to see that a palindrome with an even number of digits is divisible by 11 is as follows. Take one, such as 314413 and split it into parts:

$$314413 = 300003 + 10010 + 4400.$$

The last term is clearly a multiple of 11. So is 10010, because $10010 = 11110 - 1100$, and each of those numbers is a multiple of 11. Finally, $300003 = 333333 - 33330$, which is another difference of two multiples of 11. All of the parts into which 314413 was divided are multiples of 11, so 314413 is also.

In general, any palindrome with an even number of digits can be written as a sum of integers of the form $n00\ldots0n00\ldots0$ where there is an even number of 0s between the ns. In the example, 314413 is the sum of three such integers, with 4, 2, and 0 zeros separating the ns. But integers of that form can be written as a difference,

$$n00\ldots0n00\ldots0 = nn\ldots n00\ldots0 - nn\ldots n00\ldots0,$$

where there is an even number of ns in each integer on the right—two fewer in the second term than in the first. Since the number of ns is even, each integer is a multiple of 11, as $555555550000 = 11\cdot 50505050000$.

2. For the dice game, suppose that six players play the game at a dollar a go, each betting on a different number. (This is the long-term average. But even if players tend to bet more often on 6 than 1—they in fact do—it makes no difference.) If a triple occurs, the house takes in \$5 and pays out \$3 for a profit of \$2. If a double occurs, the take is \$4 from the four losers and the payoff is \$3 to the two winners, for a profit of \$1. If all three dice show different numbers, the house collects \$3 and pays \$3, for no gain.

Three dice can fall in $6 \cdot 6 \cdot 6 = 216$ ways. In the long run, each way will occur equally often. To make the calculation easy, we will suppose that we toss the dice 216 times and each way occurs once.

The number of ways they can fall and show three different faces is $6 \cdot 5 \cdot 4 = 120$ since the first can show any number, the second can be different from the first in five ways, and the third can be different from the first two in 4 ways. These give no profit to the house.

Only six triples can occur. These give the house a profit of $12.

Thus there are $216 - 120 - 6 = 90$ ways that three dice can fall so that two faces are the same. In each of these the house's profit is $1, for a total of $90.

So, in our 216 games a total of $216 \cdot 6 = 1296$ dollars is bet and the house's profit is $90 + 12 = 102$ dollars. That is, the house edge is $102/1296 = 17/216$ or 7.9%.

For those familiar with probability, this can be seen in another way. The player's outcomes are

Gain	Probability
3	$(1/6)^3$
2	$3(1/6)^2(5/6)$
1	$3(1/6)(5/6)^2$
-1	$(5/6)^3$

So, the expected gain per play is

$$3\left(\frac{1}{6}\right)^3 + 6\left(\frac{1}{6}\right)^2\left(\frac{5}{6}\right) + 3\left(\frac{1}{6}\right)\left(\frac{5}{6}\right)^2 - \left(\frac{5}{6}\right)^3 = -\frac{17}{216},$$

as before.

In effect, the house returns to you, on the average, $1 - (17/216)$ of what you bet, or 92.1 cents for each dollar. It does not take too many repetitions of this transaction to reduce the size of a bankroll, and enough repetitions will reduce it to zero. In the meantime, the house will be inexorably raking in money: if there are only our six players, betting $5 a round (dollar games are scarce even in casinos of the lower class) at the leisurely pace of one round per minute, the average profit to the house is in excess of $100 an hour.

References for further reading

Martin Gardner, *More Mathematical Puzzles and Diversions*, Penguin Books, 1966, Chapter 13, "James Hugh Riley Shows, Inc.".

John Scarne, *Scarne's New Complete Guide to Gambling*, Simon & Schuster, 1974, Chapter 19, "Casino Side Games".

CHAPTER 6

The U.S.—Iraq war

We were at the Robin Hill House restaurant, above Cobh.

"The prix fixe dinner is only € 38, Owen," Richard had said. "You can get a decent bottle of Australian red for €45. All that is before tip, but surely that's a small enough price to pay for material for an outstanding column. Not that I in any way view your hospitality as payment, of course."

A counteroffer of fish and chips in Cork Harbor probably would not have succeeded, so I didn't make one.

"Ah, very good," Richard said on looking at the menu. "Pan-fried saddle of rabbit with Murphy's stout, honey and prune sauce. You don't see that in the U.S. very much. Owen, what's your position on leprechauns?"

One has to be prepared for lurches when talking with Richard.

"Why, above them, I suppose," I said. "I'm taller."

"Yes, yes," Richard replied. "But are you for them or against them?"

"I've never thought about it," I said. "Surely they're harmless."

"Owen, you almost make me believe in the stereotype of the feckless Irish," Richard said. "You don't mind that a symbol of the Irish should be no more then three feet tall, dress in green, smoke a pipe, and be a trickster? You don't mind that the Irish should be thought of as *cute*? That is, when they're not thought of as being stupid, drunk, and violent?"

"It's all good for the tourist trade," I said. "Except maybe the violence."

"It's clear that Ireland is too good to be left to the Irish," Richard said. "Did you know that at this very moment there is a video camera in Ballyseanrath monitoring a fairy circle in case a leprechaun should pop up? Don't think that there haven't been plenty of reports of sightings, because there have been. A municipality in County Kerry has misguidedly put up a sign in a traffic circle, "Leprechaun Crossing". Tourists should be allowed to come and leave money, as long as they do it quietly, but they shouldn't be *encouraged*."

"Richard, you are exciting yourself too much," I said in a soothing manner. "Tell me, how did you and Michelle get together?"

"You know," Richard said, "I really can't remember. I was doing some menial work for highly insufficient pay for an organization whose name I won't mention because it deserves not to be thought of by decent people, Michelle was in the same place, and we somehow drifted together."

"It's strange how these things happen," I said.

"It is indeed," he answered. "As you may have noticed, women and men are different and one of the differences is that women tend to be less fascinated than men are by the sort of curiosities that we find so appealing. Depending on your point of view, this is either a defect in women's brains, or one in ours. Of course there are many men left cold by, for instance,

$$1 + 5 + 8 + 12 = 2 + 3 + 10 + 11$$
$$1^2 + 5^2 + 8^2 + 12^2 = 2^2 + 3^2 + 10^2 + 11^2$$
$$1^3 + 5^3 + 8^3 + 12^3 = 2^3 + 3^3 + 10^3 + 11^3.$$

'So what?' they say. 'Who cares? What's on TV?' Michelle, though, found them interesting and some of that interest transferred itself to me."

"Lucky for you," I said.

"Yes, she's one in a million," Richard said. "Humans are almost infinitely variable, you know, though it doesn't show on the outside. Two arms, two legs, a head—I'm sure that the average space alien wouldn't be able to tell one of us from another. But inside people there are whole orders of magnitude of differences that you'd never know about just by looking. There are mathematicians who can see, instantly, things that I can't and never would. They can see things that I can't see even after they're explained to me, slowly, using small words. They are *different*. And better."

I had never seen Richard so humble. I thought of telling him that inside he was several orders of magnitude ahead of the rest of us when it came to numbers, but I didn't. His present state needed to be encouraged and its opposite didn't.

"Michelle and I were clearly made for each other," Richard said. "When you change words into numbers with $a = 1, b = 2, c = 3$, and so on, *Michelle* is 67, the same as *Stein*. And *Richard* is 61, the same as *Smit*."

"Amazing," I said, and then did some calculation. "*Owen* is 57 and *O'Shea* is 48. There doesn't seem to be anything striking there."

"You're right," Richard said. "Some things are not striking."

Later we got around to the U.S.—Iraq conflict.

"The war in Iraq has thrown up some nice numerical oddities," Richard said. "I spotted some curiosities involving palindromic numbers that you and your readers may find interesting. The first U.S.-led war on Iraq commenced and ended in 1991, which was the last palindromic year of the century. Then 11 years later the coalition forces of the U.S. and the U.K. renewed the pressure on Iraq, which resulted in the United Nations passing Resolution 1441 in the 11th month of 2002. There we have four palindromes: 1991, 11, 1441, and 2002.

"The coalition forces invaded Iraq in the third month of the third year of this century. This may be written as 3/03, which is palindromic. (The war actually began on March 19, 2003. Perhaps planning took longer than expected, since had it started on March 9 the start would have been on a palindromic date, 3/09/03. I won't say that all of the difficulties in Iraq came from this first error.)

"When Baghdad fell to the coalition forces the U.S. military authorities called a press conference on April 11, announcing that they had drawn up a list of 55 of the most wanted Iraqi leaders. They went on to say that they had issued thousands of decks of cards to U.S. troops, each deck consisting of 55 cards bearing their names and photographs. Of course, 11 and 55 are palindromes and, what is more, April 11 is the 101st day of the year."

"All these palindromes," I said, "make me want to think of some pun involving 'backward' and 'forward'."

"Resist the temptation, Owen," Richard said. "You may also note that the statue of Saddam was toppled in Baghdad on April 9, 2003, at 1555 hours (British Summer Time). That date was the 99th day of the year. Your readers might enjoy finding just which days in the year are palindromic both in their month date and their position in the year, for regular and for leap years.

"President Bush declared an official end to hostilities on Thursday, May 1, 2003. That happens to be the 121st day of the year, which is yet another palindrome. The largest gathering of world leaders after the announcement that hostilities had ceased in Iraq took place in St. Petersburg, Russia on Saturday, May 31, 2003. That is the 151st day of the year. Note that 151 is a palindrome."

"5/31/03 isn't palindromic," I said, "but if you add 53103 to its reversal 30135 you get 83238, which is a palindrome. Amazing!"

"No, I'm afraid it isn't," Richard said. "It happens much of the time: $123 + 321 = 444$, $2005 + 5002 = 7007$."

"Oh, all right," I said. I immediately thought of $89 + 98 = 187$, but decided to keep my thoughts to myself.

"U.S. forces killed Saddam Hussein's two sons on July 22, 2003," Richard went on. "The 101st Airborne Division carried out that U.S. attack. Note that

22 and 101 are both palindromic. The U.S. Government had offered a reward of 15 million U.S. dollars for information leading to the capture, dead or alive, of each of Saddam Hussein's two sons, Uday and Qusay. The U.S. Secretary of State, Colin Powell, approved that reward, totalling $30 million, on July 31, 2003. That happens to be the 212th day of the year. I probably do not need to say it, but I will anyway: 212 is another palindrome.

"Consider the number of the U.N. Resolution: 1441. Partition this palindromic number as 14 41. The sum of 14 and 41 is 55, the number of playing cards in the deck of most wanted Iraqis.

"Consider 14 41 again. The sum of the last two digits is 5, and the number of combinations of 5 objects chosen from 14, $\binom{14}{5} = \dfrac{14 \cdot 13 \cdot 12 \cdot 11 \cdot 10}{5 \cdot 4 \cdot 3 \cdot 2 \cdot 1} =$ 2002, the palindromic year U.N. Resolution 1441 was passed. The sum of fourteen consecutive primes, starting from 107 up to 179, is 2002. The sum of the numbers from 1 to 55—the number of most wanted Iraqis—is 1540. Here are four curious equations involving 1441 and 55.

$$(41^2 - 14^2) + 55 = 1540 = 1 + 2 + 3 + \cdots + 54 + 55$$
$$55 \cdot (41 - 14) + (14 + 41) = 1 + 2 + 3 + \cdots + 54 + 55$$
$$(144 - 1)(5 + 5) + 55 + 55 = 1 + 2 + 3 + \cdots + 54 + 55$$
$$1441 + 1441 + 144 - 1 = 55^2$$

"The last U.N. Resolution prior to the war on Iraq was passed on November 8, 2002. It is curious that precisely 131 (another palindrome) days passed before the war commenced on March 19, 2003. Multiply 131 by 11 (November is the 11th month) and you get 1441, the number of the now famous resolution. It was of prime importance in world affairs, so it may not be surprising that it is linked with the first three primes, 2, 3, and 5: in fact, $2! \cdot 3! \cdot 5! + 1 = 1441$."

"Richard, I would have thought that puns on *prime* would be beneath you," I said.

"I was thinking of your readers," he said. "They may want to play the game of representing 1441 using the increasing and decreasing sequences of the nine digits and inserting plus, minus, times, and divides in appropriate places."

Solutions will be given at the end of the chapter.

"Incidentally, that 11-year span at the end of the last millennium between the two palindromic years, 1991 and 2002, is interesting. At the end of the previous millennium there was only a two-year span between the two palindromic years, 999 and 1001. At the end of the next millennium there will be an eleven-year span between the two palindromic years, 2992 and 3003."

"And it's back to two years from 9999 to 10001," I said.

"We probably won't live to experience it," Richard said. "The name *Saddam* is not palindromic. But Saddam's first name may in some curious way reflect his split personality. Split his first name in two parts as follows: Sad dam. The first part reads *sad*. The second part in reverse reads *mad*.

"Saddam was captured on 12/13/03. Partition this number in two halves: 121 303. Each half is palindromic, continuing the palindromic trend. The Iraqi regime under Saddam's rule has been described as part of an axis of evil, so it is no surprise that $121 \cdot 303 = 36663$, a palindromic number that contains within it the notorious number of the beast.

"Saddam was 66 years old when he was captured, a number incidentally that is also palindromic. Here is a curious equation connecting the numbers 1441 and 66.

$$1441 + 1441 + 1441 + (44 - 11) = 66^2.\text{"}$$

"Wouldn't it be better if that last summand was $(14 - 41)$?" I asked.

"Yes," said the professor. "but the sum then would be $66^2 - 66 + 6$, not quite as nice. You can't have everything in life, Owen. You should know that by now. You may like some of the following identities better:

$36663/66 = 555.5$,
$(36663 \cdot 1441)/(55 \cdot 66) = 14554.1 = 101 \cdot 144.1$,
$14^3 + 41^3 = 71665$,
$(71665/55 = 1303$, the day of the month and the year of Saddam's capture),
$41^3 - 14^3 = 66177$,
$(66177/55 = 1203.218\ldots$. The first four digits give the month and the year of Saddam's capture).

"The figures in the date of Saddam's capture, 12/13/03, can be written as 1 21 30 3, which sum to 55. Now write the date of Saddam's capture as 12/13/2003, or 12132003. $(121 \cdot 3 \cdot 2 + 0 + 0) \cdot 3 = 2178$. That number multiplied by its reversal is 18974736, which is 66^4, and there is 66, Saddam's age when captured, again. (The number 2178 is the only known number greater than 4 that when multiplied by its reversal gives a fourth power.)

"Saddam was captured in the 12th month, on the 13th day. The product of 12 and 13 is 156. The product of their reversals, 21 and 31, is 651, the reversal of 156. You may also note the following curiosity: $156 \cdot 651 = 101556 = 273 \cdot 372$. Integers that are the product of reversals in two different ways are not common.

"Consider the date of Saddam's capture: 12/13/03. Write it as 121303. Square that to get 14714417809. That contains within it the number 1441, the number of the now famous U.N. Resolution."

"And the sum of its digits is 46, only two less than the value of *O'Shea*," I said.

"True," Richard said. "Continue to practice, Owen. You may get better."

Addenda

On July 1, 2004, Saddam Hussein appeared in court in Iraq on a number of charges, including charges of war crimes. The day that Saddam appeared in court was the 202nd day he was held in custody since his capture by U.S. troops on December 13, 2003. I hardly need to add that 202 is a palindromic number.

On October 19, 2005, the 292nd day of the year, his trial began in Baghdad which, Richard pointed out, was appropriate:

$$14 \cdot 41 + 1 + 4 + 4 + 1 = 292 + 292 = 2^9 + 2 + 14(4 + 1),$$
$$2^9 \cdot 2 + 292 + (1 + 4)^{(4-1)} = 1441,$$
$$292292/1441 = 202.\ldots.$$

Day 292 is exactly 4/5 of the way through the year, or $(1 \cdot 4)/(4 + 1)$ of the way along. His trial started on day number $14 + 4 + 1$, month number $1 + 4 + 4 + 1$, in the year $1441 + 14 \cdot 41 - 1 - 4 - 4 - 1$. He was charged with, among other things, the murder of $144 - 1$ Shiites.

Solution

To represent 1441 using 123456789 and 987654321, it is possible to use only addition and multiplication to get

$$1 + 2 \cdot 3 \cdot 4 \cdot 56 + 7 + 89 = 1441$$

and only addition, subtraction and multiplication to get

$$9 \cdot 87 + 654 + 3 + 2 - 1 = 1441.$$

References for further reading

Ken Anderson, *Coincidences: Chance or Fate?* Blandford, 1995, pp. 62–63.

Martin Gardner, *Knotted Doughnuts and Other Mathematical Entertainments.* W. H. Freeman, 1986, chapter 1, "Coincidences".

Webb Garrison, *Civil War Curiosities.* Rutledge Hill Press, 1994.

CHAPTER 7

The number of the beast

I had taken Richard to Dublin so that he could do something at Trinity College. I asked him to meet me at the Dublin Zoo when he was done. I was waiting when he got off the number 26 bus.

"Well, Owen," he said upon alighting, "Why are we here?"

I told him that it would soon be made plain, paid our admission fees (recently raised to €12.50), and took him inside. "Did you know, Richard, that this is the third oldest zoo in the entire world?" I asked, not expecting any reply. His head is full of numbers, but the date of the zoo's founding, 1830, isn't a prime, or a palindrome, or a sum of two cubes in two ways, and he had no reason to attach it to a zoo. "Let's look at the animals. The ones in the African Plains may be wondering when this long cold snap is going to end and the climate will get back to normal."

After a stroll through the zoo, during which Richard noted that it was difficult to tell which was more intelligent, the residents of Monkey Island or those looking at them, we settled into the zoo's best restaurant, not much above the fast-food category but sufficient for our purposes.

"The last time we met," I said, "you mentioned some things about 666. I looked in the Bible and found Revelation 13 : 18, 'Let him that hath understanding count the number of the beast: for it is the number of a man; and his number is six hundred threescore and six.' I thought that would be a good topic for a column and I hoped you could give me some items. Looking at beasts in the zoo would be just the thing to put you in the proper mood."

"Owen, when will you stop underestimating me?" Richard said. "I don't need external stimuli. 666 has been thoroughly investigated by recreational mathematicians over the years, but I have some things that will probably be new to your readers. By the way, reams of nonsense have been written about the identity of the man whose number was 666. It's impossible to know for sure, but it's most likely that the emperor Nero was who was meant."

"I know a few things about 666," I said. "It's the sum of the numbers from 1 to 36, and $666 = 6^3 + 6^3 + 6^3 + 6 + 6 + 6$. And," I went on, trying to impress him—actually I had prepared beforehand—"it's $1^6 - 2^6 + 3^6$ and $6 + 6 + 6 \cdot (6^2 + 1^2)$."

"Very good," he said, "But you might want to check that last computation."

"Oops!" I said. "There's a typographical error. Here, let me—"

"No, don't correct it. Let your readers make it right." (The answer is at the end of the chapter.) "Well then," Richard went on, "have we come to the limits of your knowledge? I hope not, because 666 is with us all every day—the number of seconds in a day, 86400 is $6! \cdot 6!/6$. Did you know that the sum of the squares of the first seven primes is 666?"

"Indeed it is," I said, as I quickly calculated $2^2 + 3^2 + 5^2 + 7^2 + 11^2 + 13^2 + 17^2$.

"Here is symmetry:

$$1^3 + 2^3 + 3^3 + 4^3 + 5^3 + 6^3 + 5^3 + 4^3 + 3^3 + 2^3 + 1^3 = 666.$$

That's actually too pretty for a beast. Here's symmetry of a different sort:

$$7 \cdot 7 + 13 \cdot 31 + 17 \cdot 71 - 19 \cdot 91 + 23 \cdot 32 = 666."$$

"Consecutive primes, except for 11, and consecutive plus signs, except for one minus," I alertly pointed out.

"You can't have everything, Owen," the professor said. "You can, though, have these variations on the theme of 6! and 666:

$$(6!/(6 \cdot 6)) - (6 + 6)/6 = 6 + 6 + 6,$$
$$6! + 6 + 6 - 66 = 666,$$
$$(6! + 6 + 6) - (6 \cdot 6) - 6(6 - 6/6) = 666, \text{ and}$$
$$\frac{6! + 6! + 6!}{\dfrac{66}{6} - \dfrac{6}{6}} = 6 \cdot 6 \cdot 6.$$

If you're not tired of triples of 6s, $666 \cdot ((66/6) - (6 \cdot 6)/6 - (6 + 6)/6) + (66 - 6)$ is 666 in base $6 + 6 + 6$."

"Bravo!" I couldn't help saying. Richard doesn't really need to have his ego expanded, but sometimes he outdoes himself. "They remind me of the game of expressing integers using exactly four 4s, as $2 = 4/4 + 4/4$ and $3 = (4 + 4 + 4)/4$. I expect that quite a lot could be done with six 6s."

"That could be another challenge for your readers. Beginners find getting 31 difficult with four 4s, but it's not hard with six 6s: $6!/(6 \cdot 6) + 66/6$. Here's a representation with six 6s without a factorial: $31 = 6 \cdot 6 + 6 - 66/6$. But this is off the beastly topic. You probably are not aware that the number of seconds in 6 days and 6 hours, when divided by 6!, gives $6! + 6^2 - 6$."

"How could anyone possibly be *aware* of such a thing?" I asked.

"It depends," Richard said, "on your state of development. Those monkeys out there are probably not aware that 666 is the sum of two consecutive palindromic primes, 313 and 353, and the giraffes, not as quick to grasp things as monkeys, would no doubt fail to understand that 666 is a triangular number."

"It is?" I said.

"It's the 36th triangular number (more 6s!—$36 = (6 + 6 + 6) + (6 + 6 + 6)$): $1 + 2 + \cdots + 36 = (36 \cdot 37)/2 = 666$. In other bases, 666 is triangular in base 49 and 2040."

"So $6 \cdot 2040^2 + 6 \cdot 2040 + 6$ is a triangular number. Which one is it?" I asked.

"Your readers have calculators, even as you do, Owen, available for use," Richard answered.

I defended myself by saying, "I was only asking because I thought it might be something striking, like the 666th. But go on."

"Very well, but you should know by now that *all* integers are striking in one way or another. Here is a curious Pythagorean triple, $216^2 + 630^2 = 666^2$, or, to make it more striking, $(6 \cdot 6 \cdot 6)^2 + (666 - 6 \cdot 6)^2 = 666^2$. The 693, 1924, 2045 right triangle has area 666,666. It's not known if there is any other Pythagorean triangle whose area consists of all 6s (except the 3, 4, 5 triangle, of course), or, for that matter, of any one digit repeated any number of times. It's likely that there are no others. The beast, after all, is unique."

"I seem to remember, "I said, "that there is another number of the beast."

"Yes," the professor answered, "in some very early manuscripts of Revelation, its number is 616. If that is the beast's true number, then it's no longer startling that Stanley Kubrick, the director of the movie *2001*, died 666 days before January 1, 2001, but I could generate any number of equally startling facts using 616. And if 666 was no longer the number of the beast, the highway now called U.S. 491 could revert to U.S. 666, as it was until it was changed in 2003 for some reason that I suspect was silly. I'm happy that in Ireland the Lismore-Fermoy road is designated as R666, as it has been and should continue to be. But here's a different topic that your readers might like to look into: Smith numbers."

"What's a Smith number?" I asked, as I was expected to.

"A composite integer whose sum of digits is the same as the sum of the digits of its prime factors. I assume you need an example," Richard replied.

"Yes, please." I take pride in my ability to maintain my equanimity even when Richard is being annoying. Sometimes he is not doing it on purpose.

"$58 = 2 \cdot 29$ is a Smith number because $5 + 8 = 2 + 2 + 9$. The first ten Smith numbers are 4, 22, 27, 58, 85, 94, 121, 166, 202, and 265. There are infinitely many, and 666 is one of them: $666 = 2 \cdot 3 \cdot 3 \cdot 37$ and $6 + 6 + 6 = 2 + 3 + 3 + 3 + 7$. In bases other than 10 you'd get an entirely different list of Smith numbers, of course. That just occurred to me!"

Richard paused for thought, not for very long, and then went on. "I don't know if there are any binary Smiths. Smith numbers in base 2, that is. But there must be integers that are Smiths in two or more bases. I wonder if anyone has looked for them. Be that as it may, Smith numbers have a connection with primes all of whose digits are 1. Some people call them "repunit primes", which I find uncouth, but I suppose I have to use the term to make myself understood. But I will use it sparingly. It's a fact that if R_n is a prime $111 \ldots 1$ with n 1s, then $1540 R_n$ is a Smith number. So—here's an example, Owen—$1540 \cdot 11 = 16940 = 2 \cdot 2 \cdot 5 \cdot 7 \cdot 11 \cdot 11$, and we get a digit sum of 20 for the number and for its prime factors. There are multipliers other than 1540 that have the same property."

"Are we straying from the topic of the beast?" I had the temerity to ask.

"Not because there's nothing more to be said," Richard replied. "Mathematics is inexhaustible and even the beast is large, but one thing can lead to another, you know. The number of the beast has had altogether too much nonsense written about it, so it's fitting that it's related to phi, the so-called golden ratio, $\varphi = (1 + \sqrt{5})/2 = 1.61803398 \ldots$, that has had almost as much nonsense written about it."

"Yes," I said, "I think I remember reading that someone had determined that the ratio of people's heights to the distance from their navels to the ground isn't really φ, as some people—φ freaks, you could call them—had claimed. But what's the relation?"

"Here's one," said the professor. "Sin $666° = -\varphi/2$. And here's another: cos $216°$ (or cos $(6 \cdot 6 \cdot 6)°$) $= -\varphi/2$. And, speaking of φ, there's the Euler φ-function—$\varphi(n)$ is the number of positive integers less than n and relatively prime to n—and $\varphi(666) = 6 \cdot 6 \cdot 6$."

"φ brings π to mind," I said. "Is there any connection between 666 and π?"

"Certainly! Here's π to five significant figures: $0.666^6 \cdot 6 \cdot 6 = 3.141578 \ldots$. And $\dfrac{83758}{26661}$ (note the 666) is $3.14159258 \ldots$—seven significant figures. The sum of the first 144 digits of π is 666, which is fitting because $144 = (66 + 6) + (66 + 6)$. The 666th digit of π is 3 and the next two are 4 and 3, and $343 = 7 \cdot 7 \cdot 7$. The first occurrence of 666 in the decimal expansion of π is at decimal places 2441, 2442, and 2443, and $2441 + 2442 + 2443 = 11 \cdot 666$. If π is a normal number, which everyone believes but no one has been able to prove, then 666 appears infinitely often in its decimal expansion. A normal number, as you should know, Owen, is one in whose decimal representation every sequence of digits appears with its expected frequency. That is, in the long run, 1/10 of its digits are 6s, and for three-digit sequences, the frequency of 666 is, on the average, one in a thousand. In fact, 666 will appear infinitely often at decimal places whose last three digits are 666, 667, and 668. But of course if π is normal, then π knows everything."

"How is that?" I asked.

"Take *Ulysses* for example," the professor said, ascending (or descending, depending on your point of view) into the style of a lecture, "and make it into a number by any code that you want, for instance $a = 001, b = 002, \ldots,$ $z = 026$, with other numbers for capitals, punctuation marks, and other symbols. You could even include a code for the end of a line and another for the end of a paragraph. Or use ASCII. Or anything else, it doesn't matter. However you code it, *Ulysses* is turned into a sequence of digits—long, but finite. If π is normal, then that sequence of digits appears somewhere in its decimal expansion. Decode π and you will come across *Ulysses*. Infinitely often, in fact. Every other book that has ever been written is in π as well. Every book that *could* be written is there. The proof of the Riemann Conjecture is in π. π knows everything! Your biography is there, Owen, including your date of death. Tomorrow's issue of the *Irish Examiner* appears, in full. Your next column is there also. If you knew where to look, you could copy it instead of going to the trouble of writing it."

"But won't π also include all possible lies? My biography will appear with infinitely many incorrect dates of death," I said.

"Owen, you sometimes surprise me with your acuteness," Richard said. "You're right. For each correct copy of *Ulysses* there are infinitely many erroneous ones. There are infinitely many where the last period is a comma, or an exclamation point, and infinitely many attributing it to John Joyce, or to Owen O'Shea. Combing through the entrails of π is not a practical way of getting at anything valuable, even assuming that we had all of its decimal places available on demand."

"Next comes relations of 666 with e," I said.

"I'll have to disappoint you," Richard said. "I don't have any. But they are not to be expected, since e is the base of natural logarithms and the beast is highly unnatural. Similarly, you won't find any between 666 and $\sqrt{2}$ because $\sqrt{2}$ is not transcendental, whereas the beast certainly transcends anything we know. But even so, a sufficiency remains. Take the first eighteen integers, discard the primes, divide the remaining numbers into two sets, and behold:

$$4^2 + 6^2 + 8^2 + 9^2 + 10^2 + 12^2 + 15^2 = 666,$$
$$1^2 + 14^2 + 16^2 + 18^2 = 777.$$

Those who dislike 666 tend to think that 777 is a favorable number. The equations show how even-handed the integers are. Now consider the arrangement of digits on calculators and keypads:

$$7 \quad 8 \quad 9$$
$$4 \quad 5 \quad 6$$
$$1 \quad 2 \quad 3.$$

Take the digits in any row, horizontal, vertical, or diagonal, as a three-digit number. The difference between its square and the square of its reversal is a multiple of 666. For example, $741^2 - 147^2 = 527472 = 792 \cdot 666$. For another, $654^2 - 456^2 = 219780 = 330 \cdot 666$."

"What," I said, "that *always* works? Let me try 951." I punched my calculator and, sure enough, got a multiple of 666, $1320 \cdot 666$.

"It lends support to those who think that calculators and computers are the devil's work," Richard said. "Here are some more pretty 666 equations:

$$666 + 666 = 6 \cdot 6 \cdot 6 \cdot 6 + 6 \cdot 6,$$
$$666 + 666 = 66(6 + 6 + 6 + 6/6) + 66 + 6 + 6,$$
$$\frac{6 \cdot 6! + 6 \cdot 6! + 6 \cdot 6!}{6 \cdot 6 \cdot 6} = 66 - 6, \text{ and}$$
$$666 + 666 = \left(\frac{66}{6}\right)^3 + \frac{6}{6}."$$

"The 3 in the last one takes the edge off a bit," I said.

"All right, Owen, have it your way. Replace it with $\frac{6+6+6}{6}$."

It's hard to get around Richard.

"Here is a number that seems unremarkable, $666^2 - (66 + 6)^2 = 438372$. But its square root is $662.0966697\ldots$. What is the chance that of all three digit sequences, 666 should appear so early, as the third one, right after 096 and 966, in that decimal expansion?"

"Let's see," I said. "The first two aren't 666 and the third is, so the chance is

$$\frac{999}{1000} \cdot \frac{999}{1000} \cdot \frac{1}{1000} = .000998.$$

That's certainly statistically significant."

"Your reasoning is a bit short of watertight, Owen," said Richard. "But the occurrence is indeed unusual. Now, here's a fraction that seems to imply that removing beastliness does not change things very much, not a pleasant conclusion:

$$\frac{1666}{6664} = \frac{1}{4}.$$

Take the 666s away, and it's still 1/4."

"No, Richard," I said. (By chance he had hit on one of my favorite topics. It's not often that I get the chance to impress him.) "It's just a special case of the Law of Universal Cancellation, which says that if the same symbol appears twice in any expression it can be cancelled. For instance,

$$\frac{532}{931} = \frac{52}{91} \quad \text{and} \quad \frac{865}{346} = \frac{85}{34}.$$

It has nothing to do with the beast. The Law works with algebra, too. Take

$$\frac{a^2 - b^2}{a - b}.$$

Cancel an a and a b and you're left with one a and one b in the numerator. Then two minuses make a plus, so you get $a + b$."

"Ah," Richard said, not batting an eye. He wasn't going to give me any satisfaction. But I knew that I had scored. "Here, then," he went on, "are two instances of the Law of Invariance of Digits

$$2^5 \cdot 9^2 = 2592 \quad \text{and} \quad \frac{666}{64676} = \frac{6 \cdot 6 \cdot 6}{6 \cdot 46 \cdot 76}$$

and one of the Law of Persistence of Patterns

$$66 + 6 = 12 \cdot 6,$$
$$666 + 66 + 6 = 123 \cdot 6,$$
$$6666 + 666 + 66 + 6 = 1234 \cdot 6,$$
$$66666 + 6666 + 666 + 66 + 6 = 12345 \cdot 6,$$

and so on. You could challenge your readers to show why the pattern holds. It's not difficult to see. It's almost as easy as 1, 2, 3. Speaking of which,

$$\frac{2^{13}(13 - 2) - 2^{13} - (1 + 3)/2}{123} = 666.$$

Since the sum (and product) of 1, 2, and 3 is 6, $111 + 222 + 333 = 666$. Ask your readers to find other such equalities. Some are obvious, like $123 + 232 + 311$ and $332 + 223 + 111$, but not many would find $321 + 312 + 31 + 2$."

"So whether they find the hard one or not they'll be happy," I said. "That's just the sort of thing that I need."

"Do you sometimes feel at sixes and sevens, Owen? That's a curious phrase, by the way. It goes back to Chaucer, but how it came to mean a state of confusion isn't clear. Perhaps it reflects the difficulty that people had with multiplication back then when the numbers got large, like 6 times 7. In any event, multiply 7776 by 6667. ($7776 = 6 \cdot 6 \cdot 6 \cdot 6 \cdot 6$, by the way.) The product is 51842592. Note the 2592 that we saw a minute ago—you can't keep a good number down. If you are sensitive to integers, you'll also note that $5184 = 2 \cdot 2592$. Now divide by 6 to get 8640432. You'll no doubt now immediately see that $864 = 2 \cdot 432$. Divide by 6 again. The quotient is 1440072 and I need hardly say that $144 = 2 \cdot 72$."

I got out my calculator. I think Richard appreciates slack-jawed amazement, but I wanted to show him that I was capable of more than that. "Yes," I said, "and another division by 6 gives 240012, the same property again, and another gives 40002, the same thing *again*. Could such a thing happen with any other numbers?"

Richard said, "I'll leave it to your readers to determine just how astonishing that example is. Something I wouldn't leave to them is the task of representing 666 as a fraction whose numerator contains each of the digits $1, 2, \ldots, 9$ exactly once and whose denominator is a palindrome. In fact, I'd hesitate to ask them to do that even leaving off the condition about palindromes. Those with rudimentary programming skills could no doubt do it, but in these days when you can buy programs to do almost everything, the number of people who can dash off a quick program to solve a little problem is declining. Progress does have its costs. Why, I can remember the days when it mattered, when you were writing code, what the first letter of a variable was, and—"

"Yes, yes," I put in. Reminiscences are tedious, except for mine. "What's the fraction?"

"There are two, and only two:

$$\frac{913572846}{1371731} \quad \text{and} \quad \frac{264197538}{396693}.$$

The denominator of the first is prime, by the way. Those numerators bring to mind another puzzle: insert some plus signs in appropriate places in the sequence 123456789 to make the sum 666, and do the same for 987654321. There are two solutions for the first but only one for the second."

"Richard, you've given me more than enough material for one or even two columns. I'm very grateful."

"Wait, Owen, don't rush off," Richard said, "I have a few more items that might be of interest."

"Good," I said as I settled back. I really had had enough for one day, but Richard is a valuable source and mustn't be offended. Also, I doubt if anyone else will listen to him on the subject of numbers. We must do our best to be kind to our fellow creatures.

"Square 666 and you have 443556, and its cube is 295408296. Do you notice that the lucky number 7 does not appear? Noteworthy, I'd say. But more surprising is that $4^3 + 4^3 + 3^3 + 5^3 + 5^3 + 6^3$ plus the sum of the digits in 666^3 is 666. The only other number with this property is 2583."

"Does 2583 have any sinister significance?" I asked.

"I'll leave that to your sinister-minded readers," he said. "Did you know that when Nelson Mandela was a prisoner his prison number was 46664? That's palindromic, contains 666, and is $4 \cdot 11666$. What's more, $46664/5714 =$ $8.166608333\ldots$, a number, call it m for short, that contains 666 and has the property that

$$2m = 16.33321666\ldots, \quad 4m = 32.66643332\ldots, \text{and}$$
$$8m = 65.33286664\ldots.$$

They all contain 666. And—don't interrupt, Owen, I know that $10m$ has a 666 as well—

$$46664 \cdot 5714 = 266638096, \quad \frac{266638096}{4} = 66659524, \text{ and}$$

$$\frac{66659524}{4} = 16664881.$$

Also, 666 in binary is 1010011010. Write that in reverse, then replace every 0 with a 1 and every 1 with a 0, and the result is 666 in binary once again. The smallest prime containing 666 is 6661."

"Stop!" I said. "I'm drowning in 666s."

"All right," Richard said, "Let's look at another number. You've heard of the fine structure constant in physics? It's a dimensionless constant and so its value is independent of what units we use to measure things. The acceleration of gravity is constant too, but it's 9.8 if we measure in meters and 32 if we measure in feet, not to mention measuring time in seconds. The value of the fine structure constant is different—it's built into nature. Its reciprocal is very close to 137, and some people thought that it was exactly 137. Richard Feynman said—I have the quote right here,

It has been a mystery ever since it was discovered more than fifty years ago, and all good theoretical physicists put this number up on their wall and worry about it. Immediately you would like to know where this number for a coupling comes from: is it related to π or perhaps to the base of natural logarithms? Nobody knows. It's one of the greatest damn mysteries of physics: a magic number that comes to us with no understanding by man.

The value is close to 137.036, and I've read that it's actually one of those variable constants. At high energy levels its value can be closer to 128. (Energy levels of something or other, don't ask me what. I've never understood physics. "Field," "friction, "force"—what do those f-words *really* mean? I can't get my mind around them. They're not like numbers. Numbers are rock-solid, dependable and not slippery.) 137 is a number like 666 that has had nonsense written about it, though the amount is small compared with that of beast-nonsense. Consider this: $1^3 + 3^3 + 7^3 = 371$, a cyclic permutation of digits. Also, $-2^0 + 2^1 + 2^3 + 2^7 = 137$. In binary, 137 is 10001001. Take that number in base 10 and divide it by 137. Go ahead, Owen, do it."

"All right," I said, pulling out my calculator. "It's 73000.007. Is that significant?"

"Think more deeply than your calculator, Owen. It's 73000 with remainder 1. Read that backwards—100037. We started with 137, divided by 137, and got 137 again. The number really is built into the universe. While you have your calculator out, enter a four-digit number."

"Any one?" I asked.

"Any one. Now multiply it by 137, and then multiply the result by 73."

"Well," I said, "look at that!"

"Yes, it's impressive for people up to the age of, say, 13.7. People older than that, and your readers, should be able to explain the phenomenon. While we are in the neighborhood of 137, let's stop in on 153, the seventeenth triangular number and the number in the miraculous draft of fishes mentioned in Luke's gospel."

"I know," I said, "you're going to bring up $1^3 + 5^3 + 3^3 = 153$, the same thing that happens with 371."

"Yes," the professor said, "and you probably know that $4^3 + 0^3 + 7^3 = 407$. But $153 = 1! + 2! + 3! + 4! + 5!$. And 153 is related to 666: add the six permutations of 1, 3, 5:

$$153 + 351 + 135 + 531 + 315 + 513 = 3 \cdot 666.$$

Here are the permutations again, in mirror image:

$$153 + 315 + 531 = 135 + 513 + 351.$$

"But there's nothing special about that, that'll always happen whenever—" I started to say.

"Note also," Richard interrupted, "that $153 = 3 \cdot 51$, something that seldom happens, and that $1^0 + 5^1 + 3^2 = 1 \cdot 5 \cdot 3$. Changing the subject, but not by very much, from 153 to 163, did you know that

$$e^{\pi\sqrt{163}} = 262537412640768743.99999999999925\ldots$$

is almost an integer?"

"What a coincidence!" I said.

"Actually, it's not a coincidence," the professor told me. "It happens because the imaginary quadratic field with discriminant -163 has class number 1, but we don't need to go into that. You can solve that equation for π, with the term on the right rounded off, to get

$$\pi \approx \frac{\ln(640320^3 + 744)}{\sqrt{163}},$$

accurate to 30 places. Now that π has come up—this zoo coffee is unusually strong, my mind is racing—I'm reminded of some more relations, some only close approximations, of course,

$$\varphi^{1.37} = \frac{\pi}{\varphi}, \quad e^{1/\pi} = 1.37, \quad 3.71^2 = 13.7,$$

$$\ln\left(\frac{137^3}{\varphi^2}\right) = 13.7, \quad (e\pi)^{-2} = .0137."$$

"I am amazed. Is *everything* related?" I asked.

"That is a deep question, Owen," Richard said, "best left to the philosophers, though seeing what they are up to nowadays they will probably not consider it. Here are two final 666 equations:

$$123 + 1^2 + 2^2 + 3^2 + 1 \cdot 23^2 = 666; \quad \frac{1 + 2 + 3}{6} = \frac{1 \cdot 2 \cdot 3}{6} = \frac{\sqrt{1^3 + 2^3 + 3^3}}{6}$$

By the way, if you like to think that computers are beastly devices, here is support: if you assign numbers to letters by putting $a = 6, b = 12, c = 18$ and so on, the sum of the letters in *computer* is 666."

"Calculators are still safe, I hope," I said as I put mine away. It was almost time for the restaurant to close. "Incidentally, did you see that a new Mersenne prime has just been found?"

"Of course, Owen," he answered. "I was one of the first to know. It's

$$2^{25964951} - 1,$$

7,816,230 digits long, the forty-second of its kind. If computers are inventions of the devil, then he must want us to find out new things. Before the days of computers, the largest known prime was $2^{127} - 1$, a puny thing with fewer than 40 digits. By the time you tell your readers about the new prime, an even bigger one may well have been discovered. Many computers are at this moment, and every other moment, whirring away in the Great Internet Mersenne Prime Search and a bigger one may turn up before you put that one in your column. But no matter how many we find, we'll be no nearer a proof that there are infinitely many Mersenne primes. Everyone believes that's true, but we probably won't live to see a proof.

"For $2^n - 1$ to be prime, n must be composite. A challenge for your readers would be for them to show that if n is even, then $2^n - 1$ is composite. I expect that some of them could rise to it. Fewer would be able to show that $2^n - 1$ is composite when n is odd and composite, but some might. Be sure to tell them that if they do, they are very good indeed."

"All right, Richard," I said, "I will. Let's be on our way and leave all the beasts in peace."

Solutions

1. The incorrect $666, 6 + 6 + 6 \cdot (6^2 + 1)$ should be $(6 + 6 + 6) \cdot (6^2 + 1)$.

2. To see why the pattern

$$66 + 6 = 12 \cdot 6,$$
$$666 + 66 + 6 = 123 \cdot 6,$$
$$6666 + 666 + 66 + 6 = 1234 \cdot 6,$$
$$66666 + 6666 + 666 + 66 + 6 = 12345 \cdot 6,$$

persists (at least for a while), divide everything by 6 and it should become clear.

3. A difficult-to-find representation of 666 using only 1s, 2s, and 3s is

$$666 = 132 + 23^2 + 3 + 1 + 1.$$

Representations such as $666 = 1 + 1 + \cdots + 1$ (666 terms) will not receive any extra credit.

4. Ways of inserting $+$ signs in 1234567889 and 987654321 to get 666 are

$$1 + 2 + 3 + 4 + 567 + 89,$$
$$123 + 456 + 78 + 9, \text{ and}$$
$$9 + 87 + 6 + 543 + 21.$$

5. One way to show that $2^n - 1$ is composite if n is even is to write $2^n - 1$ in binary notation, which is $111\ldots11$, all 1s, n of them. If n is even, then there are $n/2$ blocks of 11, so that $2^n - 1$ is divisible, in binary, by 11, and hence by 3 in decimal. Another way is to note that

$$2^{2k+2} - 1 = 4(2^{2k} - 1) + 3\,.$$

So if $2^{2k} - 1$ is divisible by 3, then $2^{2k+2} - 1$ is as well, since both terms on the right of that last equation are multiples of 3. Since $2^2 - 1$ is divisible by 3, it follows that $2^4 - 1$, $2^6 - 1$, $2^8 - 1$, $\ldots$ will also be divisible by 3.

If you hit on the idea of writing $2^n - 1$ in binary (not likely unless you are very clever), you can use it to show that $2^n - 1$ is composite whenever n is composite. If n is composite, then it's divisible by something other than 1 or n, call it k. Then the n 1s in the binary representation of $2^n - 1$ can be divided into k blocks of n/k 1s. Thus $2^n - 1$ will be divisible by the number whose binary representation is $111\ldots11$ (n/k 1s). For example, in binary, $2^{15} - 1 = 11111, 11111, 11111$ and so it is divisible by binary 11111, or 31. In decimal, $2^{15} - 1 = 32767 = 31 \cdot 1057$.

Or, if you are skilled in algebra, you can note that

$$2^{kr} - 1 = (2^r - 1)(2^{(k-1)r} + 2^{(k-2)r} + \cdots + 1)\,,$$

which shows that $2^n - 1$ is composite when n is.

Addendum

Since this chapter was written, a new largest Mersenne prime has been discovered. I will not mention what it is because by the time you read this, a yet larger one may well have been found.

References for further reading

Paul Hoffman, *Archimedes' Revenge*, Penguin Books, 1991, Chapter 1.

Underwood Dudley, *Numerology*, Mathematical Association of America, 1997, Chapter 7.

Martin Gardner, *The New Age*, Prometheus Books, 1988, Chapter 23.

CHAPTER 8

Curios of the Lusitania and other curious matters

Richard, Michelle, and I were having dinner in Cobh, my home town. Richard inquired about its name.

"*Cobh* is the Gaelic name for cove," I said. "It is pronounced the same as 'cove'. Cobh was a transatlantic port of call for many years. The vast majority of those who emigrated from Ireland in the nineteenth century during the Great Famine left from Cobh. Later, if you wish, I will take you to the Clonmel Churchyard (locally known as The Old Church Graveyard) just outside of Cobh. The Lusitania victims are laid to rest there. The poet Reverend Charles Wolfe, who wrote 'The Burial of Sir John Moore,' is buried there. So also is the well-known Irish heavyweight boxer, Jack Doyle."

"Most interesting," said Richard. "Let me see. First, the Lusitania. The sinking of that ship, with the great loss of life, is a well-known tragedy. Second, I know of Wolfe's famous poem. Third, I have read about Jack Doyle, who was known in his day as 'The Gorgeous Gael'.

"I even know that Cobh was called Queenstown until 1922, when its name reverted to Irish, but that the River Lee has not been translated to An Laoi. Why do Irish rivers have such short names—the Moy, the Tar, the Nore, the Deel? I suppose it's because they're such short rivers. The English are even worse—the Dee, the Exe, the Wye. You'd think that they'd have more imagination than to name their rivers after letters of the alphabet. Now where I come from rivers are rivers: the Choctawhatchee, the Apalachicola, the Waccasassa. The U.S. has its faults, but being constricted is not one of them."

"I would," I said, "certainly agree with you on that point. Incidentally Richard, my editor is very pleased with the information that you are giving me for my column. He said that your curiosities will definitely attract new readers, as well as helping me to hold on to the ones I have! He is a curious old bird, you know. Only the other day he was telling me that in his childhood he lived in a house numbered thirteen, which contained seven rooms. He then gave me the following curiosity: $(7 \cdot 13 + 7 + 13) \cdot (-7 + 13) = 666$, which is the famous number of the beast! I was wondering if you had any curious information on those two primes. If my editor likes the info you give it might help to keep the old man off my back for a while!"

"Those two numbers," said Richard, "are popular numbers, even with non-mathematicians. The number 7 is usually considered a lucky number, while of course, quite the contrary may be said of 13. In the western world 13 is considered to be the unluckiest number of all. Sevens, of course, are everywhere. There are seven days in a week, seven seas, seven continents, and Seven Wonders of the World. Having made the world, God rested, we are told, on the seventh day. There are seven deadly sins, and seven heavenly virtues. A child reaches the age of reason, we are told, at seven years. Since Biblical times we are told that the average lifespan on this earth is seven decades. Shakespeare wrote of The Seven Ages of Man. To be extremely happy is to be in seventh heaven. To randomize a deck of 52 cards at least seven shuffles are needed. Those are all well known. Not everyone, though, has noticed that the seventh day of the seventh month of the seventh year of the seventh millennium will fall (under the Gregorian calendar) on the seventh day of the week."

"I certainly didn't," I said.

"The number 7 has many interesting mathematical properties," said Richard. "For example, the only known values of n for which $n! + 1$ is a square are 4, 5, and 7. It is curious that $7! + 1 = 71^2$. It's obvious that $n! + 1 = (10n + 1)^2$ has no other solutions, but it might be interesting to look at variations on that theme. In another direction, $(2^{p-1} - 1)/p$ is a square only when p is 3 or 7. Also, if a and b are the two short sides of an integral right-angled triangle, then one of a, b, $a - b$, and $a + b$ is evenly divisible by 7."

"My," I said. "Could more than one of them be a multiple of 7?"

"Come, Owen, use your head. If any two of them are divisible by 7 then clearly all of them are. That means that the triangle has sides $7A$, $7B$, and $7C$ and is only a magnification by a factor of seven of another right-angled triangle. Such triangles are as dull as . . . dishwater."

He caught himself before he could say, "as dull you are today." Richard is not the soul of tact, but he is better than he could be.

"Consider the integers 1741725 and 7776. Except for the large number of 7s among their digits, they seem unassuming and ordinary, do they not? But

$$1741725 = 1^7 + 7^7 + 4^7 + 1^7 + 7^7 + 2^7 + 5^7.$$

It is not ordinary that 7776 is a perfect power, 6^5. Let's look at 7776^5. It is 28,430,288,029,929,701,376. Partition it into five groups of four-digit numbers: 2843, 0288, 0299, 2970, and 1376. What is their sum?"

It's annoying to be asked questions when you know that the questioner knows the answer, but Richard deserves to be humoured, and after all, I did ask him for some ideas. After writing the numbers down and getting out my calculator I was able to say, "7776."

"Quite right," he said. "Of course, you have probably noticed that $7^3 = 343$, or $343 = (3 + 4)^3$."

"I never thought of it that way," I said. "But going back to 7776, it made me think—666 is the number of the beast, and I remember that some people have called 777 the number of God. Does 777 have any interesting properties?"

"You should know by now, Owen, that *all* integers are interesting, though some are more interesting than others. Don't you remember the proof of the theorem 'n is interesting, $n = 1, 2, 3, \ldots$'?"

"Yes," I said. "Suppose that there are integers that are not interesting. Then there is a smallest one, call it N."

"But that makes N quite interesting, contradicting the assumption that it was not and proving the theorem," Richard said quickly, not waiting for me to complete the proof, which might have taken longer. "777 needs only one digit for its representation in base 10, and the same is true in base 6, where it is 3333. Also, 777 has a link to 13:

$$\frac{131313}{13 \cdot 13} = 777.$$

Further, $7! + 7! + 7! - 1$ is evenly divisible by 13.

"Consider the story," Richard went on, "of the Lusitania, the giant liner that was sunk by the Germans off the south coast of Ireland in 1915. Seven seemed to crop up quite a lot in its history. The Lusitania was launched on June 7, 1906. She sailed from Liverpool to New York on her maiden voyage on September 7, 1907. The Lusitania was 787 feet in length. You may not have noticed that $787 = 777 + 77/7 - 7/7$. The Lusitania had a Gross Register Tonnage of 31,550. The sum of those digits is twice 7. She had a horsepower of 70,000. The Lusitania commenced what was to be her last voyage when she left New York on Saturday, May 1, 1915. Note that the words *New York* contain seven letters and that Saturday is the seventh day of the week. Germany was the nation responsible for sinking the Lusitania. The initial letter of *Germany* is the seventh letter of the alphabet, and the word *Germany* contains seven letters. The captain of the German U-Boat (a man named Walter Schwieger) responsible for sinking the Lusitania estimated in his logbook that it was 700 meters from his submarine at the time the torpedo was fired."

"The Lusitania," said the professor, "went down at 2:10 pm. Note that 210 is evenly divisible by 7. The Lusitania went down off the Old Head of Kinsale, on the south coast of Ireland. About ten miles west is a landmark known as Seven Heads Point. The Lusitania passed Seven Heads Point shortly before it was torpedoed. Incidentally, the master of the Lusitania on its last voyage was Captain William Thomas Turner. He captained both the Lusitania and its sister ship, the Mauritania, after becoming a Cunard captain in 1907. Using the usual code $a = 1, b = 2$, and so on, the sum of the letters in *Lusitania* is 106, whose digits sum to 7. What's more, the sum for *Cunard* is 61, whose digits also sum to 7."

"There are a lot of sevens there," I said.

"There's more," said Richard. "The Lusitania required a crew of 77 able seamen, but only 41 were aboard on its last voyage. The number of American citizens lost on the Lusitania was 128, or 2^7. It was sunk on the 127th day of the year, and $127 = -1 + 2^7$. Finally, the Lusitania went down on Friday, May 7, 1915."

"Seven did not appear to be lucky for the Lusitania," I said.

"Nor for those aboard her," Richard said. "Let us leave the sea, but not 7. Pretty things abound, as $(7 - 1)! + 1 = 721$, whose reversal, 127, is $-1 + 2^7$. The number of minutes in a week is $2 \cdot 7!$ Your readers may also be interested to know that the largest number of eclipses, both solar and lunar, that can occur in one year is seven. This last happened in 1982, when there were four of the sun and three of the moon. (All three lunar eclipses were total.) This will not happen again until the year 2485, a year that is a multiple of 7."

"What about 13?" I enquired. "It's almost twice 7. What else does it have going for it?"

"Quite a bit" said the professor. "Did you ever notice that $-1^3 + 3^3$ is twice 13? Or that $13^2 = 169$, and its reversal, $31^2 = 961$, which is the reversal of 169? The digits of 13, or 31, sum to 4, and the digits of 169, or 961, which are the squares of 13 and 31 respectively, sum to 16, which is the square of 4. What's more, $169 = (16 + 9) + (16 \cdot 9)$. The 13th prime number is 41, and $2 + 3 + 5 + 7 + 11 + 13 = 41$. Also, the sum of the first thirteen prime numbers is 238 and the sum of the digits of 238 is 13."

"And the remainder when 238 is divided by 13 is $1 + 3$," I said.

"Very good, Owen," he replied. "Do you remember Wilson's theorem?"

"Well—maybe you should state it," I said.

"It's very memorable, Owen," Richard answered. "It says that $(n - 1)! + 1$ is divisible by n when, and only when, n is prime. So, since $(7 - 1)! + 1 = 6 \cdot 5 \cdot 4 \cdot 3 \cdot 2 \cdot 1 + 1 = 720 + 1 = 721$ and 721 is divisible by 7, we know that 7 is prime. On the other hand, $(6 - 1)! + 1 = 120 + 1 = 121$ is not divisible by 6, so 6 is not a prime. Since 13 is prime, $(13 - 1)! + 1$ is divisible by 13."

"Yes," I said, working my calculator. "$12! + 1 = 36846277 \cdot 13$. But that seems rather a lot of trouble to go to just to find that 13 is a prime."

"Beauty needs no practical justification," the professor said. "Did you know that on rare occasions $(p - 1)! + 1$ is divisible not only by p, but by p^2? In fact, there are only three primes less than $5 \cdot 10^8$ for which that is true, and one of them is 13. The other two are 5 and 563. Here's a curiosity: add the fifth powers of the first 13 integers

$$1^5 + 2^5 + 3^5 + 4^5 + 5^5 + 6^5 + 7^5 + 8^5 + 9^5$$
$$+ 10^5 + 11^5 + 12^5 + 13^5 = 1002001.$$

The sum is palindromic and is $7^2 \cdot 11^2 \cdot 13^2$. The smallest prime whose reversal is a prime different from itself is 13. Speaking of reversals, in 2013 Easter falls on March 31 and in 2031 Easter falls on April 13. The smallest prime that can be expressed as the sum of two primes $(2 + 11)$ and two composites $(4 + 9)$ in only one way is 13. The 31st prime is 127, and $-1 + 2 \cdot 7 = 13$. Also, $(13 - 3 + 1)^{1 \cdot 3} = 1331$. The first two digits in $13^{\cdot 13}$ are 1 and 3."

"Stop, Richard! I give up! You and 13 win," I said. "Let's move on to 91, the product of 7 and 13. Anything interesting about that number?"

"What a question! 91 is $1 + 2 + 3 + \ldots + 13$, the 13th triangular number. It is also $1^2 + 2^2 + 3^2 + 4^2 + 5^2 + 6^2$, $3^3 + 4^3$, and $6^3 - 5^3$. Also, $91 = (9^3 - 1^3)/(9 - 1)$. Or, to make that more striking at the cost of a little artificiality,

$$91 = \frac{9^{\sqrt{9 \cdot 1}} - 1^{\sqrt{9 \cdot 1}}}{9 - 1}.\text{"}$$

"What about Friday the 13th?" I asked. "I know that there is at least one every year."

"That is correct," said the professor. "Some years have two Friday the 13ths, and some have three. Three is the maximum."

"How often do they occur?" asked Michelle.

"As you probably know," said Richard, "the present calendar repeats itself every 400 years. In that period there are 97 leap year days, so the total number of days is 146,097, which is 20,871 weeks exactly. In those 20,871 weeks Friday falls on the 13th of the month 688 times. Therefore the probability that any Friday, chosen at random, falls on the 13th of the month is $688/20871 = .03296. \ldots$ That means that there is about one chance in 30 that a Friday picked at random will be a Friday the 13th."

"Which is the most likely day of the week to fall on the 13th?" I asked.

"Believe it or not," said the professor, "the answer is Friday."

"Gosh! What is the most unlikely day of the week to fall on the 13th?"

"The most unlikely days of the week to fall on the 13th are Thursdays and Saturdays," said the professor. "The 13th of the month falls on a Thursday 684 times every 400 years. It falls on a Saturday the same number of times. Mondays and Tuesdays do slightly better. The 13th of the month falls on each of those days 685 times in a 400-year period. It occurs on Sundays and Wednesdays 687 times every 400 years. Here is a little card I have prepared. Take one."

He then handed me a card, titled "Curiosities Concerning Friday the 13th" that contained

1. Two successive years can each contain one Friday the 13th only, but three successive years cannot do so.

2. Four successive years can each contain two Friday the 13ths, but five successive years cannot do so.

3. A leap year can contain three Friday the 13ths, but this happens only fifteen times in a 400-year period.

4. Two successive years cannot contain three Friday the 13ths.

5. The most likely day of the week for the 13th of the month to fall on is Friday.

6. When three Friday the 13ths occur in a non-leap year, the previous year will contain just one Friday the 13th (and that will always be in June) while the following year will also contain just one Friday the 13th, which will occur in either May or August.

7. When a leap year contains three Friday the 13ths, the following year will contain two, and those will always be in September and December.

8. A Friday the 13th occurs in February in a leap year just 13 times in a 400-year period.

"Leap years rarely have three Friday the 13ths," said Richard. "In fact, it happened only three times in the last century, in 1928, 1956, and 1984. It happens every 28 years if and only if the last year of a century is a leap year. The year 2000 was a leap year, therefore the rule applies, and so 2012 will contain three Friday the 13ths. So also will 2040, 2068 and 2096. The rule then breaks down, because 2100 will not be a leap year. Perhaps your readers, Owen, can calculate the next leap year beyond 2096 that contains three Friday the 13ths."

"I'll give them the chance," I said. "Have you any other oddities concerning Friday the 13th?" I asked.

"Many, but I don't have the time to give them all. I will give just two more snippets of information. There is about one chance in 212 that a child will be born on a Friday the 13th. You may also be interested to know that the thirteenth day of a year such as 131313, or say 13131313, where the year number consists of at least three sets of 13, is a Friday the 13th under the Gregorian calendar."

"I doubt that our calendar will last that long," I said. I saw Richard starting to react, but I beat him to it. "I know, being practical interferes with beauty. Are there any famous people connected to 13, Richard?"

"Well," said the professor, "we spoke earlier of the heavyweight boxer Jack Doyle, who was famous in Ireland, England, and the U.S. in his day. He never became a champion, mainly because of his fondness for wine, women, and song. He was also a compulsive gambler. He once said that he lost his fortune because of fast women and slow horses. He spent his final days living rough on the streets of London. It was all very sad. Anyway, there was a string of 13s that cropped up in his life, although he and everybody else were apparently unaware of it."

"Give me some examples, please, Richard." I was eager to find out about this, as I had heard of Jack Doyle since the days of my childhood. He was born in Cobh— which I never tire of identifying as my home town —in a district known as The Holy Ground. Doyle had been one of my heroes since boyhood. He was the man who had claimed he could sing like John McCormack (a famous Irish tenor) and box like Jack Dempsey. Later in life I learned that some wit had allegedly said (rather unkindly and unjustly, I thought) that Doyle could sing like Jack Dempsey and fight like John McCormack.

"Jack Doyle was born on August 31 (note the reversal of 13) in 1913," the professor said. "There was a solar eclipse on that day. He took size 13 shoes. His first airplane flight occurred on Friday, January 13, 1933. He bought a house for his parents at 65 College Road, in Cork city. Note that 65 equals 5 times 13. He had 23 professional fights, and won 17 of these, 13 by knockouts. Doyle was 65 years and 104 days old when he died in the 78th year of the century. Of course, 65, 104 and 78 are all multiples of 13. The sum of those three numbers, $65 + 104 + 78$ is 247, which has a digit sum of 13. Doyle died on December 13, 1978."

"The number 13 is the sixth prime," I said. "Have you any unusual information concerning those first six primes?"

"Well," said Richard, "the numbers 788, 789, 790, 791, 792 and 793 are divisible by 2, 3, 5, 7, 11 and 13 respectively."

"But 794 is unfortunately not divisible by 17," I said.

"True," said the professor. "But you can remedy that defect. Do you recall our friend 777? Can you find a multiple of 17 that leaves a remainder of 777 when it's divided by $2 \cdot 3 \cdot 5 \cdot 7 \cdot 11 \cdot 13 = 30030$?"

"Well, not right at the moment," I had to confess.

"If you could," the professor said, "you would find that $N = 210997$ is the smallest integer such that $N + k$ is divisible by the kth prime for $k = 1, 2, \ldots, 6$."

"What?" I said.

"I'll spell it out, Owen. 210998 is divisible by 2, 210999 by 3, 211000 by 5, 211001 by 7, 211002 by 11, 211003 by 13, and 211004 by 17."

"But 211005—" I began, but thought better of it. To get something that would be divisible by 19, Richard would no doubt bring up

$$2 \cdot 3 \cdot 5 \cdot 7 \cdot 11 \cdot 13 \cdot 17 = 510510$$

and he might actually try to make me do something with it.

"Curiously," said the professor, "13 is the *fifth* lucky number, the *sixth* prime number and the *seventh* Fibonacci number."

Richard, with characteristic modesty, assumed that because he knew what the lucky numbers were, everyone must know. This may not be the case. There are several types of numbers which mathematicians often refer to as "lucky numbers". One type is those commonly referred to as the "lucky numbers of Euler". A second set of integers (those that Richard was referring to) that are termed "lucky numbers" can be generated as follows: Begin with the sequence of odd integers:

$$1, 3, 5, 7, 9, 11, 13, 15, 17, 19, 21, 23, 25, 27, 29,$$
$$31, 33, 35, 37, 39, 41, 43, \ldots.$$

The first number after 1 is 3, so cross out every third number, leaving

$$1, 3, 7, 9, 13, 15, 19, 21, 25, 27, 31, 33, 37, 39, 43, \ldots.$$

The first number after 3 is 7, so cross out every seventh number:

$$1, 3, 7, 9, 13, 15, 21, 25, 27, 31, 33, 37, 43, \ldots.$$

Continue in this way. When you are finished, the surviving numbers—they were lucky not to be crossed out—

$$1, 3, 7, 9, 13, 15, 21, 25, 31, 33, 37, 43, 49, 51, 63, 69, \ldots$$

are the lucky numbers. Though they are not all prime, they share many properties of the prime numbers. For example, all even integers seem to be sums of two lucky numbers.

"Also," Richard went on, "the sum of the remainders when 13 is divided by all the primes up to 13 is 13. The 13th Fibonacci number, 233, is also a prime.

The successive digits 1 and 3 appear for the first time at the 111th place in the decimal expansion of π. Note that 111 equals $3 \cdot 37$ and that the digits 3, 3 and 7 sum to 13.

"The number 13 is also important in the so-called 'real world.' I find the world of mathematics considerably more real than that of, say, rock stars, but most people seem to think otherwise. There were 13 people around the table at the Last Supper. There were 13 original states in the U.S., there are 13 loaves of bread in a baker's dozen, 13 weeks in a quarter of a year, and 13 cards of any one suit in a normal deck of playing cards. Do you recall the Apollo 13 space trip to the moon in April, 1970?"

"Yes, that was the trip that almost ended in disaster," I said.

"That's right," said Richard. "Apollo 13 took off from Pad 39 (note the multiple of 13) on Saturday, April 11, 1970. The three astronauts aboard were James Lovell, Jack Swigert, and Fred Haise. The total number of letters in the three names is 31, the reverse of 13. When they lifted off from the Kennedy Space Center the clocks at Mission Control in Houston were reading 1313 hours. Two days later, on Monday, April 13, an oxygen tank exploded in Apollo's service module, resulting in the systems in the service and command modules being crippled. The astronauts were brought back to earth safely, but the whole trip came to within a hair's breadth of disaster. Curiously, this ill-fated Apollo trip splashed down in the Pacific on April 17, 1970 at 13 hours, 07 minutes, and 41 seconds precisely. The number 130741 is evenly divisible by thirteen. The Mission Designation of Apollo 13 was AS-508. The digits of 508 sum to thirteen. It was strange that this Apollo trip, ominously named Apollo 13, should have come so close to ending in tragedy."

After dinner I brought the professor and Michelle to Clonmel Churchyard, that old and historic burial ground just outside Cobh. As we drove the short journey there Richard offered me the following proposition bet:

"Let us check the license plates of the next twenty cars we see," he said. "Note the last two digits of each number. I will bet you €20 at even money that at least two of the next twenty cars we see will have plate numbers in which the last two digits are identical."

I thought for a moment about the bet. The last two digits of each number could be anything from 00 to 99. There are 100 different possibilities to be considered. I reasoned that the chance any two are the same are 1/100. The chance that two are the same with twenty cars involved is therefore 20/100. Thus, there must be 80/100, or four chances in five, that the last two digits in two registration numbers will not be the same.

"The bet sounds good, Richard, but I know better than to bet with you. How did Damon Runyon put it? 'One of these days, a guy is going to show you a brand-new deck of cards on which the seal is not yet broken. Then this guy is going to offer to bet you that he can make the jack of spades jump out of this brand-new deck of cards and squirt cider in your ear. But do not accept this bet, because as sure as you stand there, you will wind up with an ear full of cider.' Let's just make a hypothetical bet."

We commenced noting the numbers of cars we happened to see. The last two digits of the twelfth and sixteenth cars were identical.

"You win a hypothetical €20," I said, "But I bet I'll win the next hypothetical bet."

"How much?" asked Richard.

It happened again! There were two cars with identical last two digits well before twenty cars.

"What am I missing?" I asked. "Or are you just extraordinarily lucky today?"

The professor said that he was no luckier than usual and then explained how to calculate the true odds involved in this proposition bet. The reader may find it a pleasant exercise to figure the true odds. The solution is given at the end of this chapter.

When we reached the churchyard all three of us paid our respects to the Lusitania victims, to the Reverend Charles Wolfe, and last but by no means least, to the legendary Jack Doyle.

"How is it," I asked Richard, "that you know about Jack Doyle?"

"I've always been interested in the sport of boxing and its practitioners," he said. "Boxing is in a sad state of decline these days, but it may some day regain some of the popularity it had when there were boxing matches on television almost every night of the week and championship bouts could sell out baseball stadiums. Joyce Carol Oates appreciates boxing, and A. J. Leibling wrote superbly about it. I think that it causes less physical damage than football and fewer deaths than automobile racing. You know that Jack Doyle was married to Movita, the Mexican actress who later married Marlon Brando?"

"Yes, indeed," I said. "Jack Doyle is still remembered in Ireland. He was almost the British heavyweight champion. Jack Doyle's Bar is still going strong in Cobh. Let's go to the Holy Ground district—I'll show you where he was born and raised."

I recounted to Michelle and Richard how the body of Doyle had been brought home from London in 1978 and how the people of Ireland had given him a hero's funeral.

"On that bleak December day," I said, "people from all walks of life, from every corner of Ireland, and many also from abroad, walked behind the hearse carrying Jack's body. All undoubtedly recognised, that Jack—like all of us—had had many faults in life. But most, if not all, also appreciated the great qualities of the man and were deeply saddened by the cruel change of circumstances Jack had endured in life. They knew that this old warrior had met in life 'with Triumph and Disaster, and had treated those two impostors just the same,' as Rudyard Kipling had put it in his famous poem 'If,' a poem incidentally that Jack had loved. It was this feature of Jack's character, his cheerfulness even when the tide of fortune had turned against him, that endeared him to all who knew him and to all who knew of him."

As we stood there in the Holy Ground the professor and Michelle gazed silently out at the opening of Cork Harbor. We could see a ship in the distance leaving the old port. All was quiet, except for the sound of the seagulls' cry. It appeared to me that the professor's mind was wandering back in time to the glory days of the Holy Ground and to the Gorgeous Gael who grew up there.

Two days later I received an e-mail from Richard. This is what he wrote:

Dear Owen.

I hope you are well.

Michelle and I were deeply moved the other day as we stood in the quiet Holy Ground district of Cobh and heard you speak of the legendary Jack Doyle. I happen to like poetry so I was very pleased when you referred to Kipling's famous poem "If." Last night I penned a few words about the life of Jack Doyle. I think these few lines sum up his life and reflect the feelings that many Irish people still have for him. I believe many older people in the U.S. may remember Jack also. I hope you like the poem. If you wish to publish it, please feel free to do so.

Here is the poem.

Jack Doyle
by
Richard Stein

The Holy Ground is quiet now,
Unlike the days of yore.
When Jack Doyle grew up there,
A fighter to the fore.

They say he could have been World Champ,
The greatest of them all.

But wine and women, and the good times,
Led to his downfall.

We recall the fights of other days,
The punches, hard and fast.
The promise of a World Crown,
Alas. It never came to pass.

The days when you were young, Jack,
When you could do no wrong.
You were the Boxing Master,
You could sing an Irish song.

Many came to claim you, Jack.
They tried to put you down.
But they miscalculated,
The boy from the Holy Ground.

There were plenty of contenders,
No shortage to be found.
But one huge punch from Jack,
Sent them sprawling to the ground.

Life was good and sweet then,
You had friends, far and near.
But when the hard times came,
Did many of them care?

You were a hard hitter,
Of that there is no doubt.
And life hit you some cruel blows,
But could never knock you out!

Many a day you were down,
Friends were scarce and few.
You never threw the towel in,
You were a fighter, through and through.

Because of that we love you, Jack,
You were our joy and pride.
The day you left this Earth,
A living legend died!

Although you're dead and gone, Jack,
You're still remembered in Cobh town.
The boy who won fame and fortune,
The boy from the Holy Ground.

Addendum

There are only four seven-digit numbers that equal the seventh powers of their digits:

1,741,725
4,210,818
9,800,817
9,926,315

Solutions

1. The first leap year after 2096 that contains three Friday the 13ths is 2108, when they will occur in January, April, and July.

2. Here's how to calculate the odds in the professor's proposition bet: The chance that the last two digits in the first two of the plate numbers are not identical is 99/100. The chance that the last two digits of the number of a third car is different from the first two is (99/100)(98/100). The chance that the number of a fourth car differs from the first three is (99/100)(98/100)(97/100). And so on. For twenty cars, the chance of no match works out to be 0.1304, or about 13%. So, about 87% of the time, there will be two cars with identical final two digits. Even if the professor offered odds of five to one on this bet, the proposition would have still favored him. The reader may like to see just how many (or how few!) cars it takes to make this an odds-on bet.

References for further reading

William T. Bailey, Friday the thirteenth, *Mathematics Teacher* **62** (1969), 363–364.

John Haigh, *Taking Chances*, Oxford University Press, 1999.

Michael Taub, *Jack Doyle*, Stanley Paul, 1990.

CHAPTER **9**

Wordplay and other curiosities

The call was from Richard.

"Owen, I have some things to tell you," he said. "Perhaps we could meet for dinner tonight in Cork?"

"Of course," I said. Even though I would as usual be paying, what Richard has to say would be worth the cost, short notice or not.

"Excellent," he said. "Michelle may be coming as well. May I suggest the PI Restaurant on Washington Street? I thought it would be appropriate, even though the initials probably stand for something like 'Pirandello Italian' and have nothing to do with π. But they serve pizzas, circular ones, so there's a chance that they do. One mustn't underestimate the range of knowledge of restaurateurs. They do extensive work on ranges."

I could hear him stifling a chuckle at his cleverness. He was clearly pleased with himself and with the prospect of a good meal, so I didn't correct him and say that "line cook" would be more accurate.

When we settled in at the restaurant, Richard assumed the air of a host. "Look, Michelle," he said while looking at the menu, "here's a pizza with duck confit, aubergine (they mean eggplant), and roast garlic. And here's another one with black pudding, leeks, and apple sauce. I'll bet no one ever thought of that in Firenze."

After we ordered, Richard got down to business.

"Have you noticed the pattern in Irish cities, Owen?" he asked.

"What pattern? The streets are crooked enough."

"No, no, Owen. You think on too small a scale. Look at the big picture. Consider the town nearest the center of Ireland," said Richard. "It's named Athlone. The largest city in the northern half of Ireland is Belfast, the largest city in the southern half is Cork, and the largest city in the east is Dublin. A, B, C, D."

"Too bad," I said, "that the largest city in the west half of Ireland is Galway."

"Yes," said Richard. "But there's Ennis in the southwest. I expect it to grow explosively."

"You obviously like patterns, Richard."

"Mathematicians always look for pattern and order," said Richard. "Even outside of mathematics. Perhaps that is why many mathematicians enjoy playing with words. Of course words are made up of letters, which are akin to the symbols of mathematics, and mathematicians are attracted to chess, a game that is essentially manipulating symbols, so it's really all one. Skilled chess players can see things that duffers can't, and mathematicians are always asking, 'Do you *see* that?' It's all about looking and seeing. Here's a pattern that I saw. You know that there were four U.S. presidents, Garfield, Kennedy, Lincoln, and McKinley who were assassinated. K, L, and M are in alphabetical order. If we note that Garfield's first name was James, we have J K L M."

"That's a bit of a stretch. It's too bad that reality sometimes refuses to order itself properly," I said. "I suppose that when wordplay is mentioned, most people think of anagrams. Like *eleven plus two* and *twelve plus one*." I said, proud of myself for having that available. "That's a good one, isn't it?"

"Yes," said Richard. "A good anagram should have some connection with the original word or sentence. For instance, *a stitch in time saves nine* can be anagrammed to *this is meant as incentive*, a nice connection. On the other hand, anagramming *Owen O'Shea* to *no shoe awe* would be good only if it were well known that you were not impressed by footwear. And making *Richard Stein* into *It's a rich nerd* is doubly false."

"That's right, Richard," I said, "You're not rich."

"Or a nerd," Michelle loyally added.

"Indeed not," said Richard. "I suppose that *rent Irish cad* would apply were I to tear some disagreeable citizen of Ireland to pieces, but I haven't done that. You're not a cad, are you, Owen?"

"No more than you're a nerd," I answered.

"Of course not. Before getting back to mathematical topics, do you know what to say to someone who claims that 'Elvis lives'? '*Seen alive? Sorry, pal.*'"

"Why would I say that?" I asked.

"Oh, Owen," the professor sighed, "Didn't you recognize an anagram of *Elvis Aaron Presley?*"

"Ha!" I exclaimed. "Anagram that may be, but the King's name was 'Elvis Aron Presley' with only one *a* in 'Aron'."

"So it is thought in some uninformed circles," Richard answered. "He may have had that name when he was young and ignorant, but later he added the second *a*, and 'Aaron' with two *a*s appears on his tombstone."

It is *very* hard to get ahead of Richard. But I made one more try. "Well, how about an anagram of his original name?" I asked.

"What, you expect one on the spur of the moment?" Richard said. "Very well, give me a sheet of that paper you're using to take notes and give me a few moments."

He wrote, stared, wrote again and stared longer. Then there was another bout of scribbling. I thought that I had succeeded in stumping him. Then I saw his face light up—he had one. Foiled again!

"You know, Owen," he said, fairly oozing satisfaction, "many disapproving people thought that Elvis was primitive, a throwback making unseemly noises. *Reversional yelps*, in fact."

"You're wonderful, Richard," Michelle put in. I suppose she had a point.

"Here are two other nice anagrams with a mathematical flavor," Richard said. "*Gosh! See that triangle = It has got three angles* and *A decimal point = I'm a dot in place.* Speaking of dots brings the Morse code to mind: *the Morse Code = here come dots.* I suppose that with the decline in telegraphy you'll have to tell your readers that the Morse Code is a system that codes letters and symbols as dots and dashes: *e* is one dot, *a* is dot dash, or 'dit dah' as it used to be said, and *z* is dah dah dit dit. Zeros and 1s, actually, but Morse lived before the days of computers. Samuel Morse died in 1872 in New York City, too soon to have seen the *Statue of Liberty*. If he had, he would have seen that it was *built to stay free*."

It was only later that I realized that he had given me another anagram. I told the professor that my readers often like to try their hand at word puzzles, and asked if he could give me a few for my column in *The Mathematical Universe*.

"Of course," he said without hesitation. He then gave me these problems.

(1) There is a three-letter word in the English language that when spelled backwards means the same as the original word. What is it?

(2) What is special about the word *nymphly*?

(3) What word contains all five vowels in alphabetical order?

(4) What common English word, containing six letters, has opposite meanings, depending on which way it is pronounced?

I scribbled down the questions as quickly as I could. The answers to all four questions are given at the end of the chapter.

"Did you know," Richard said "that *four* has been called the only honest number in English, because it correctly states the number of letters in its name?"

"Yes, Richard, I knew that," I was able to say truthfully. "Of course, we are talking here about number names only, and not operational phrases associated with them. For example, *five squared minus five* equals 20, and there are twenty letters in that phrase, but if you don't allow such descriptions, then four is indeed the only honest number in the English language."

"Gaelic is better," said the professor, "because there are two honest numbers—dó and trí—two and three. Going back to English, *forty* appears to be the only number which has its letters in alphabetical order. I read somewhere that the smallest integer whose name uses the five vowels *a, e, i, o, u,* in any order is *one thousand five*. (Though people say 'one thousand and five', the 'and' is not part of the number name. It is a mere syntactical convenience and should be ignored.) Perhaps your readers can find the smallest number whose name uses the five vowels in order."

"I'll give them the chance," I said. The answer is given at the end of the chapter.

"Your readers may also be interested to know that in writing the names of the natural numbers in sequence, one, two, three, four, five, etc., the letter *a* is not used until *one thousand*, and *b* does not appear until *one billion*."

"How about *c*?" I asked. "I've started counting but haven't got to it yet."

"Keep at it. You're sure to find it sooner or later, certainly by the time you count up to *one octillion*."

Over the course of the evening, the professor discussed various topics. One curious story concerned Lee Harvey Oswald. Apparently Alek J. Hidell was a pseudonym that he used. It has been suggested that this pseudonym was a play on the names Dr. Jekyll and Mr. Hyde, from Robert Louis Stevenson's famous story. Some psychologists have suggested that Oswald gave himself this pseudonym, either consciously or unconsciously, as his way of letting the world know of his split personality.

"Have you any more curiosities concerning words?" I asked.

"Hundreds," he answered. "I don't have my files on wordplay with me, but perhaps I can recall some material that I believe your readers will find interesting."

The professor then went on to tell me that *ambidextrously* is the longest word in the English language that he knew of in more or less common use with no letter appearing more than once. The longest pair of non-scientific English words, he said, that are anagrams of each other, are *conservationalists* and *conversationalists*. The professor said that there was not much mixing of letters in those anagrams so that he preferred the anagram pair *undefinability* and *unidentifiably*. The words *cares* and *princes* are plural, he said, but become singular when an *s* is added at the end of either. He went on to say that the longest word he knew of that can be changed from a plural to a singular in this way was *multimillionaires*. He also said that the word *dreamt* was the only common English word that he knew of ending in *mt*.

It was good that he didn't have his files, because more and more came out of his memory. The two words *angry* and *hungry*, he said, are the only common words in the English language ending in *gry*. The word *understudy* contains the alphabetical sequence *rstu*, and he said it must be one of the few words containing four consecutive letters of the alphabet in sequence. *Weaponmaking* is one of the few words containing a four-letter alphabetical sequence (*mnop*) backwards. *Goddessship*, *headmistressship*, and *crosssection* were three of the few words each containing a triple letter. He pointed out that the invented word *Amerikkkan* was created to symbolize the racist aspect of American society because the three *k*s it contains are the initials of the Ku Klux Klan. The word *hydroxyzine*, which refers to a prescription drug, is the only word known to him, he said, containing in order the three consecutive letters *xyz*. The word *subbookkeeper*, he said, is the only word known to him that has four pairs of double letters in a row.

"I can beat that, Richard," I said. (I had heard that one before.) "Our moon is filled with many nooks and crannies, so someone who takes care of them is a *moonnookkeeper*. Five pairs!"

"A word that corresponds to nothing in external reality," Richard said loftily, "is not a word."

I could have argued that point, quite cogently I thought, but I chose not to.

"Incidentally, you probably know," said Richard, "that *yob*, a slang word for a lout, is *boy* written backwards. Well, 'mho' is the unit of electrical conductance. Conductance is the reciprocal of resistance, which is measured in ohms. Did you ever notice that *mho* is *ohm* written backwards?"

I evidently had paid more attention in my physics classes than Richard had, so I knew that that was an intentional coinage. But the hour was getting late, and Michelle's eyes had been looking glazed for some time.

"Time is passing," I said. "We'll have to leave soon."

"True," said Richard. Irrepressibly, he went on. "Speaking of time, did you ever notice that many of the time intervals we use on our clocks and in our calendars are equal to the sum of two squares? For instance, there are 3600 seconds in one hour and $3600 = 48^2 + 36^2$. There are 1440 minutes in a day, and $12^2 + 36^2 = 1440$. There are 365 days in a year—$365 = 13^2 + 14^2$. The 52 weeks in a year is $4^2 + 6^2$. There are $10 = 1^2 + 3^2$ years in a decade, $100 = 6^2 + 8^2$ years in a century, and $1000 = 10^2 + 30^2$ years in a millennium."

"But isn't everything a sum of two squares?" I asked.

"Hardly," Richard said. "The only integers that are the sums of two squares are those that do not have a prime that leaves the remainder 3 when divided by 4 raised to an odd power in their representations as a product of powers of primes."

"Oh," I said. I would have to think about that later.

"Here are even more," he went on. "Friday the 13th is said to be unlucky. $13 = 2^2 + 3^2$. One of the most special days of the year for millions of people around the world is Christmas Day, December 25th. $25 = 3^2 + 4^2$. Lent lasts 40 days. $40 = 2^2 + 6^2$. St. Patrick's Day falls on March 17, and 17 is obviously a sum of two squares. In a non-leap year July 4th is the 185th day of the year, and $185 = 4^2 + 13^2$. (Clearly, the 4 represents the fourth of July, while the 13 represents the original thirteen colonies.) The calendar repeats after $400 = 12^2 + 16^2$ years. In one 400-year period there are 97 leap days, and $97 = 4^2 + 9^2$. The only month that ever contains exactly 29 days is February, and $29 = 2^2 + 5^2$. The expression '24/7' for 'all the time' that has come into vogue in the past few years does not specify a time interval itself so, naturally, 247 is not the sum of two squares. It is, of course, the difference of two squares, namely $16^2 - 3^2$. The reversal of $2 \cdot 47$, is $49 = 2 + 47$, but I don't see how to interpret that. I can interpret $247^{0.247} = 3.89938527\ldots$ though: making 24/7 more powerful results in 52/7: 52 weeks a year, 7 days a week, likewise more powerful. Confirming that is $724^{0.724} = 117.611752\ldots$: rearrange 24/7 and there 52/7 is, also rearranged."

"Ah, here is the check," I said, pulling out my wallet.

"You are probably aware that curious events have been known to occur on curious dates," Richard went on, paying no attention. "Did you know that on February 4 $(2 + 2)$ in the year '22 (1922) 22 policemen were killed in a riot at Chauri Chaura in India? Or that on 8/8/88 the UN Secretary General announced that Iran and Iraq had agreed a truce after eight years of war? One of the most famous dates in military history is June 6, 1944, D-Day. It can be written as 6/6/44, or simply as 6644. The invasion commenced very close to six a.m. You can see the pattern here. The invasion—probably the most famous military operation ever—began very close to the sixth hour, on the sixth day of the sixth month. Three sixes! Reverse the date 6644 to obtain 4466. Is it not curious that 44/66 equals $0.666\,666\,666\,666\ldots$? Consider the time 20:02

hours, on February 20, 2002. We Americans would write that as 20:02, 02/20, 2002, a palindrome. The first time that occurred was at one minute after 10:00 on January 10, 1001, (10:01, 01/10, 1001) and it happened again at 11:11 on November 11 in 1111. Ask your readers if they can find the next such palindromic date."

"I will," I said, "but now—"

"And give your readers this one. Suppose you are given an odd number, say 73, and are asked to find two consecutive squares that differ by 73. Divide 73 by 2 to get 36.5. Then add and subtract $1/2$ to get 37 and 36. Then $37^2 - 36^2 = 73$. The procedure works for any odd number. Suppose however that you are asked to find two consecutive *cubes* whose difference is, say, 1,292,977. What is the quickest method of finding the solution?"

It is given at the end of the chapter.

"Richard," I said, getting up, "It really is time to go."

"Indeed it is," he agreed. "Time nor tide waits for no man, as they say. *Summer, autumn, winter, spring. Time's running past, we murmur.* Come, Michelle, we have things to do. We must do this again, Owen."

With that the professor waved goodbye, and was gone within the blink of an eye. It wasn't until later that I realized that he had parted leaving an anagram behind him.

Solutions

1. A three-letter word in the English language that when spelled backwards means the same as the original word is *tap*.

2. *Nymphly* is a seven-letter word containing none of a, e, i, o, and u.

3. *Facetious* contains the vowels in order. If you include y, *facetiously* satisfies the condition.

4. The word is *resign*. If pronounced one way it means, "to quit". If pronounced another way it means "to sign again" or *not* to quit.

5. The smallest integer whose name uses the five vowels a, e, i, o, u, in order is 1084, *one thousand eighty-four*. Note that it was not specified that those were the *only* vowels in the number name.

6. The next palindromic time and date to occur will be at 21:12 p.m. on December 21, 2112 or, in the U.S., 21:12 12/21 2112. In Europe it would be written as 21:12 21/12 2112. In either form the date is palindromic.

7. The quickest method I know of solving this type of problem is as follows:

Subtract 1 from 1,292,977, and divide by 3, obtaining 430,992. Take the square root of that and you get 656.4998.... Ignore the decimal part: one of the numbers is then 656 and the other is 657, and a calculator quickly shows that $657^3 - 656^3 = 1292977$. To see that this works in general, let x be the smaller of the two consecutive cubes whose difference, d, was given. Then

$$d = (x + 1)^3 - x^3 = (x^3 + 3x^2 + 3x + 1) - x^3 = 3x^2 + 3x + 1.$$

So,

$$\frac{d - 1}{3} = \frac{3x^2 + 3x + 1 - 1}{3} = \frac{3x^2 + 3x}{3} = x^2 + x.$$

The square root of this will be a number between x and $x + 1$ because

$$x^2 < x^2 + x < x^2 + 2x + 1 = (x + 1)^2.$$

So,

$$x < \sqrt{\frac{d - 1}{3}} < x + 1 \,,$$

and thus x is the integer part of the square root. Of course, you must know that d is in fact a difference of two cubes for the method to work. Not every integer has that property. If you try to apply the method to $d = 52$, for example, $\sqrt{\frac{d-1}{3}} = \sqrt{17} = 4.123\ldots$, but $5^3 - 4^3 \neq 52$.

References for further reading

William Tunstall-Pedoe and Donald C. Holmes, *Anagram Genius*. Hodder & Stoughton, 1995.

Michael Curl, *The Anagram Dictionary*. Robert Hale, 1996.

Howard W. Bergerson, *Palindromes and Anagrams*. Dover Publications, 2004.

Henry Ernest Dudeney, *300 Best Word Puzzles*. (Edited by Martin Gardner.) Charles Scribner's Sons, 1968.

C. C. Bombaugh, *Oddities and Curiosities of Words and Literature*. Dover Publications, 1961.

CHAPTER **10**

New coincidences on Lincoln and Kennedy

"Owen, you've got time for an excursion this weekend, don't you?" Richard asked.

"Oh, . . . I suppose so," I answered. I'm not adroit enough, the way that many people are, to deflect such things instinctively.

"Splendid," Richard said. "It's about 200 miles to Carrowkeel, less than six hours. We can go up one day, stay the night (at dinner I can give you some new items for your column), see the site the next day and get back before dark."

"Where's Carrowkeel?" I asked.

"I suppose you can't be expected to have all of the prehistoric sites in Ireland at your fingertips. It's a very large burial grounds, not too far from Sligo."

"Why do we want to go there?" I asked.

"Any number of reasons," Richard said. "One is that I've never been in that part of the country and neither have you, or else you wouldn't have had to ask about it. Another is that the contemplation of ancient ruins is good for the soul. But the main reason is that there's a book by an Irish author that implies that the tomb-builders had deep mathematical knowledge. It's by one of those people who can write things like, 'Stone age people had an intuitive understanding of the magnetic currents that run through the earth. They used dowsing and rock art to pass on this knowledge of metaphysics.' I want to go to see if I can feel the currents there, or find a rock with π accurate to ten places carved on it."

We went, though we had to conquer difficulties. Going down N12 near Castlebaldwin there was a sign pointing to the left—"Carrowkeel 3 km." Half a mile along there was another arrow—"Carrowkeel 5 km." After a short distance down that road was another—"Carrowkeel 6 km." Richard commented that the currents were strong, but flowing in the wrong direction. Eventually we found the site. It was impressive and would have been more so if it hadn't been drizzling, with a chill wind blowing. The tombs—rock cairns they were, some quite large—had locked gates so we couldn't have been tomb robbers even if we had wanted to. Richard pronounced himself satisfied though he had no mystical experiences, felt no currents, and saw neither UFOs nor evidence of mathematics.

The night before, I had talked Richard out of the Cromleach Lodge (much too expensive) in favor of the Yeats Tavern. I told him it was within walking distance of Yeats's grave, which turned the tide. While there he gave me some interesting information about two presidents of the U.S.

"That dreadful charlatan I. J. Matrix [see *The Magic Numbers of Dr. Matrix* by Martin Gardner—O. O'S.] gathered together some well-known coincidences concerning Abraham Lincoln and John F. Kennedy and had the nerve to cajole Martin Gardner into publishing them," the professor said. "I have some new ones. The readers of your *Mathematical Universe* column might like to read about them. They are mathematical only in the sense that they involve numbers, but they may interest your readers nevertheless. They illustrate the *flexibility* of numbers. Also, since no one planned them, they illustrate that coincidences, even a large number of them, can occur purely by chance. They will probably be insufficient to make all conspiracy theorists change their minds and abjure their theories—the tendency of the human mind to see plans, plots, and wheels within wheels, even when there are no plots, plans, or wheels anywhere, seems inborn—but if they prevent only one person from succumbing to nonsense (e.g., 'Look at all the occurrences of π in the Great Pyramid, accurate to more decimal places than the ancient Egyptians knew, it must have been built by space aliens') then they will have more than served their purpose. Conspiracy theorists should know that the elevated level of incompetence in the human race makes it improbable—impossible, even—for a conspiracy of any size to succeed or be kept secret. Why, just look at—"

"Yes, Richard," I said. He needed to be knocked off his hobby horse before it got to rocking too rapidly. "What do you have for my readers?"

"Quite a bit. First, here is a little biographical information concerning the two presidents. Abraham Lincoln was the 16th President of the U.S. He was born on Sunday, February 12, 1809. He was shot on Good Friday, April 14, 1865. He died in the early hours of the next morning. John F. Kennedy was the 35th President. He was born on Tuesday, May 29, 1917. He was assassinated on Friday, November 22, 1963.

"Here now are a number of curious facts concerning the lives of both presidents.

Lincoln was born on the 1st day of the week, in the 2nd month of the year, and died in the 4th month of the year. He finished 8th in his first attempt to be elected to public office, he eventually became the 16th President of the U.S., and it was in the year of '32 (1832) that he first ran for public office. All those numbers are in the doubling sequence; beginning with 1. The sum of 1, 2, 4, 8, 16 and 32 is 63. Those two digits are the last two digits of the year of JFK's assassination.

"Lincoln was born on the 43rd day of the year, and each of his two terms of office commenced on the 4th day of the month. $43^2 + 4^2 = 1865$, the year that Lincoln was assassinated. He was shot on April 14, during his 2nd term of office, just 5 days after the American civil war ended. It is curious that those three numbers, 14, 2 and 5, should crop up like that in such an historic occasion. Consider the number 14 raised to each of the powers of 2 and 5. Between them they contain the nine digits just once: $14^2 = 196$, $14^5 = 537824$.

"The number 11 appears to have cropped up quite a lot in JFK's life. For example, the two words *John Kennedy* contain 11 letters. The initial letter of *Kennedy* is the 11th letter of the alphabet. He was shot in the 11th month of the year, in Dallas, Texas. Those two words *Dallas, Texas,* contain 11 letters. The date of the assassination was November 22, twice 11. His car was travelling at 11 miles per hour when he was shot. He had been in office exactly 1037 days when he was assassinated, and the digits of 1037 sum to 11, as do the digits of 326, the day of the year he was shot. When Kennedy ran for election to the U.S. House of Representatives in 1946, he won the nomination in the 11th Congressional District of Massachusetts. In 1943, when Kennedy was serving with the Navy, he was assigned to patrol duty on a small boat off the Solomon Islands in the south Pacific. Shortly after midnight on August 2, 1943, a Japanese destroyer cut his boat in two. Kennedy was one of the eleven men who survived."

"Why so many 11s, Richard?" I asked. "Why not 17s?"

"The ways of numbers are passing strange, Owen," the professor said. "It is not for us to question them. Kennedy was born at 83 Beals Street, Brookline, Massachusetts. Massachusetts is the 44th largest state in the U.S. $44 = 4 \cdot 11$, $8 + 3 = 11$, and there are 11 letters in the words *Beals Street.* In the U.S. the Federal Reserve System issues Federal Reserve notes through its twelve Federal Reserve Districts. Each district is designated by a number and the corresponding letter of the alphabet. Dallas is the 11th district and is thus designated by the letter K. Many U.S. dollars, with the letter K stamped on them, and the two words *Dallas, Texas* appearing just below the K on each dollar, were placed in circulation two weeks before Kennedy was assassinated.

"On the morning of the assassination, Kennedy and his aides flew into Dallas. Soon after, Kennedy and his group left the airport and drove to the city of Dallas.

They left the airport at 11:55 am. Note the 11, and that 55 is a multiple of 11. President Lyndon Johnson took the Presidential oath (on board Air Force One) exactly 99 minutes after the assassination, another multiple of 11.

"While in office President Kennedy promised the American people that Americans would walk on the moon's surface before the end of the 1960s. On July 20, 1969, that promise was honored when two American astronauts set foot on the moon. The spacecraft that brought the astronauts to the moon was named Apollo 11.

"Kennedy was born on May 29, 1917. $2 + 9 = 11$. You may also note that the three numbers 19, 17 and 29 sum to 65. Those two digits are the last two digits of the year in which Lincoln was shot."

"And $6 + 5 = 11$," I put in.

"Precisely," Richard said. "The dates of Lincoln's shooting (written as dates are customarily written in Europe) and Kennedy's birth (written as dates are customarily written in the U.S.) have some mirror-image similarities. Lincoln was shot on April 14, or 14/4 and 144 is the square of 12, which is the day of the month Lincoln was born. (The first letter of Lincoln's surname is the 12th letter of the alphabet.) Kennedy was born on May 29, or 5/29. 529 is the square of 23. (On a 24-hour clock, 2300 hours is 11 pm.) $12 + 23 = 35$, and Kennedy was the 35th President. Also, Kennedy's age at death was $2 \cdot 23$.

"Lincoln was shot in 1865, 89 years after the founding of the U.S. Kennedy was shot in 1963, 98 years after the shooting of Lincoln. (89 reversed is 98.) Lincoln was shot on 14/4/1865 and the sum of those digits is 29. Kennedy's date of birth, 5/29/1917, contains a 29. Summing, $5 + 2 + 9 + 19 + 1 + 7 = 43$. Lincoln was born on the 43rd day of the year. Lincoln's day of birth was 12/2, and Kennedy's day of death was 11/22. Remarkably similar, are they not? Kennedy was born on the 29th of the month, and died on 11/22/1963. $29 + 11 + 22 + 19 + 63 = 144$, which represents the date Lincoln was shot. Lincoln was $56 = 14 \cdot 4$ years old when Booth shot him in Ford's Theatre. Partitioning the digits in Kennedy's assassination date as $196 + 311 + 22$ gives a sum of 529, which represents the date of Kennedy's birth.

"Lincoln's assassin was John Wilkes Booth. Kennedy's assassin was Lee Harvey Oswald. If we use the usual code $a = 1$, $b = 2$, $c = 3$, and so on, the sum of the values of the initials of both assassins is 35, and Kennedy was the 35th President of the U.S.

"Consider 1963, the year of Kennedy's assassination. $19 \cdot 63 = 1197$, a permutation of the digits of the year of his birth. Squaring Kennedy's birth day, 29, gives 841. Add 841 to 1122 (the month and day Kennedy was shot) and you have 1963, the year of the assassination.

"Subtract 1963 from its reversal: $3691 - 1963 = 1728 = 12^3$. Kennedy was shot during the 12th hour of the day. The year of Kennedy's assassination, 63, is numerically interesting, as $63 = 6^2 + 3^3$. Kennedy was shot on the 22nd day of the month in 1963. Two 2s make 4 and $4 \cdot 1963 = 7852$, an equation containing all nine digits exactly once.

"Kennedy was shot on the 22nd day of the 11th month. If you multiply, you get $22 \cdot 11 = 242$, a palindrome and a sum of four consecutive integers, $59 + 60 + 61 + 62$. (Kennedy's last full year in office was in '62.) What's more, 242 is the sum of four consecutive palindromic integers, $44 + 55 + 66 + 77$. (The 44th largest state, Massachusetts, is where Kennedy was born.) Also, $59^3 + 60^3 + 61^3 + 62^3 = 886688$, another palindrome.

"Kennedy was shot at 12:30 pm. The date of his assassination (in Europe) is usually written as 22 /11/63. $\ln(221163) = 12.30\ldots$. That number, 1230, is curious in itself. $1^7 + 2^7 + 3^7 + 0^7 = 2316$ and 2316 written in base 8 is 1230.

"Consider the equation

$$\left(24573 + \frac{6}{9}\right)(1 + 8) = 221163$$

on the left each digit is used just once and on the right is Kennedy's assassination date. Right-thinking people around the world consider the assassinations of Lincoln in 1865 and Kennedy in 1963 as wicked deeds. Of course, one of the best-known numerical symbols for evil is the number 666, which appears very early in $1963^{.1865} = 4.1127246663\ldots$.

"The lives of Lincoln and Kennedy had far-reaching effects on ordinary people around the globe. But did you ever realise that the date of Lincoln's shooting and the date of Kennedy's birth can produce two dates that have worldwide significance? $14.4 + 5.29 = 19.69$, and 1969 is the year that human beings first walked on the moon. $14.4 - 5.29 = 9.11$, and we all know about 9/11."

"Richard, I am overwhelmed," I said. "You make Dr. Matrix look like a shallow thinker."

"Well, Owen," Richard said, "I try always to present reality."

References for further reading

Martin Gardner, *The Magic Numbers of Dr. Matrix*. Prometheus Books, 1985, chapter 4, "Lincoln and Kennedy".

CHAPTER 11

Dart and card curiosities

It was evening, and Richard and I were in a bar. On the other side of the bar were a number of patrons playing darts. Richard asked me if I would like to play a game, but I declined his offer. Although I have an interest in darts I have never been a good player.

"Do you know of any good dart puzzles or curiosities?" I asked, as I pulled out my pen and notebook.

"As a matter of fact, I do," said Richard. "Here are three puzzles in darts:

(a) What is the lowest number in darts that cannot be scored with one dart?

(b) What is the lowest number in darts that one may be left where the game cannot be finished in two darts?

(c) What is the lowest number in darts that one may be left where the game cannot be finished in three darts?

"Foreign readers of your column may not be familiar with darts," Richard said. "You'll need to give some description of the game."

"That's true." I said. Since it may be that some readers of this may be in the same state of unfamiliarity, I will give enough information for them to be able to try the problems.

A dart board is circular, with its 360° circumference divided into twenty 18° sectors, labelled, in what seems to be a random order, with the numbers 1, 2 . . . 20. There is a bull's eye in the center. The sectors are divided into three regions.

If the dart sticks in one region, it scores the value of the sector. If it lands in another section, as the one labelled A in the diagram (see page 120) it scores double, and in a third, B, it scores triple. The goal of the game is to obtain a certain total, most commonly 301 or 501. Scoring starts with the total and proceeds by subtraction,

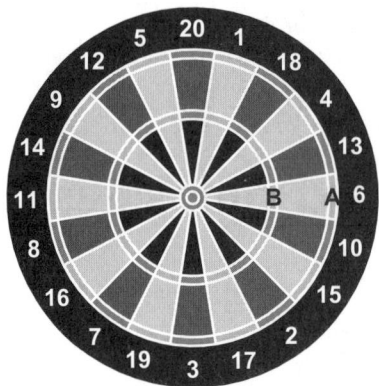

FIGURE 1

the winner being the first player to reach 0 exactly. Usually it is a requirement to start with a double, but in Ireland the game of 501 can start straight—that is, not requiring a double. Players throw a group of three darts, alternating turns. In all games, regardless of how they start, it is a requirement that the winning player finish with a double on the last dart.

"One of the oldest proposition bets concerning darts goes as follows," said Richard. "Someone challenges a good dart player to a game of darts. The hustler states that he retains any dart score he gets. If, for example, he scores 40, his score is recorded as 40. On the other hand, he informs the mark that every darts score obtained by the mark will be doubled. In other words if the mark scores 60, the score is doubled to 120. And so on."

"That seems generous," I said.

"That's the idea," said Richard, "It draws the mark in. A substantial bet is placed, and usually the cash is handed over to a neutral observer before the game. The hustler tells his mark that he does not mind if it is 301 or 501 that is played. (In fact, any starting number may be chosen, as long as it is an odd number.) The normal rules of darts apply. The only other condition is that each player begins on a double and ends on a double."

"Where is the catch?" I asked.

"Ask your readers if they can figure out why it is a very bad bet for the mark," said Richard. The solution is given at the end of the chapter. "The game of 501, straight start, can be finished in a minimum of nine darts. One way of doing this is triple 20 three times, then triple 20 three times again, then triple 20, triple 15, and double 18: $60 + 60 + 60 + 60 + 60 + 60 + 60 + 45 + 36 = 501$. If one counts the bull's-eye as a double 25, (which is the usual rule) then the game of

501, starting on a double, can also be finished in nine darts. One such way is double 20, triple 20 twice, then three triple 20s, and finish with triple 20, triple 17, bull's-eye: $40 + 60 + 60 + 60 + 60 + 60 + 60 + 51 + 50 = 501$."

I then asked, "Can 100 be scored with just three trebles?"

Quick as a flash came his reply, "Impossible." Richard then gave me a very simple proof of this. Can the reader provide such a proof?

"Incidentally, Owen," said Richard, "your readers may find it interesting to note that if one is left with 26 to finish in a game of darts, there are just 26 ways of doing so with two darts. You could ask your readers to try to find all of them."

The answer is at the end of the chapter.

"By the way," said Richard, "the twenty numbers around the dartboard can be arranged in 19! different ways, so there are 121,645,100,408,832,200 different dartboards. If you were to write a different arrangement once every minute, twenty-four hours per day, every day of the year, it would take over 231,282 million years to complete the task."

"What if I missed one through carelessness, boredom, or something?" I asked.

"Then you would just have to start over," Richard said. "To switch from one game to another, here's a card bet for your readers. The hustler asks the mark to shuffle a deck of 52 cards. He then points out to the mark that if he were to deal off a 5-card poker hand, the chance is less than $1/2$ that the five cards will contain one pair. Many poker players are aware of this. Having done this, the hustler then offers odds of 6 to 5 that he can get a pair if six cards are dealt from a shuffled deck. If the mark hesitates, the hustler can say that among the forty-seven cards still in the deck, only fifteen will match one of the five cards already dealt, so the chance of making a pair with one more card is $15/47$, which is less than a third. This explanation lies somewhere between being misleading and nonsense, but if the mark has drunk enough Guinness he will interpret it as meaning that he has a $2/3$ chance of winning and will take the bet. Actually, the odds are nearly 2 to 1 in favor of the hustler. Your readers might like to verify that."

The solution is given at the end of the chapter.

"Here's something easier. Your readers will find the answer surprising if they haven't seen it before. Take a large square sheet of paper, or cut a square out of a rectangular sheet. Suppose that the paper is just one hundredth of an inch thick. Cut the piece of paper in two halves. Place one half on top of the other half, and cut those two pieces in half so as to get four square pieces. Put them on top of one another and cut them in half, resulting in eight pieces of paper. Cut again to get sixteen square pieces and continue, each time cutting the pile in half, placing all the pieces on top of one another, and cutting in half again. After cutting the pile of paper fifty times, how high would the pile be?"

"That'll be a good easy puzzle to start a column with," I said. "I hope that readers don't write indignant letters to the editor about how difficult it is to cut pieces of paper in half while looking through a microscope."

The surprising answer to this problem is given at the end of the chapter.

"They shouldn't if you've properly primed them with the information that recreational mathematics is not meant to be practical. Speaking of primes—"

"Richard," I interrupted, "you need oil badly. That transition is the creakiest that I've heard in a long time."

"The end sometimes justifies the means, Owen," Richard said. "As I was trying to say, primes are endlessly fascinating. Is it coincidence that the answer to Douglas Adams' question about the meaning of life, the universe, and everything else was $42 = 2 \cdot 3 \cdot 7$, the second-smallest product of three primes? He said that he chose the number more or less at random, but he might not have been conscious of the primes' subtle influence on his brain. All primes except 2 and 3 are either one more or one less than a multiple of the perfect number six. There is at least one prime between any prime and its double. The primes, of course, go on forever, but we can find gaps between the primes as large as we please. For example, consider $100! + 2, 100! + 3, 100! + 4, \ldots, 100! + 100$. The first number is divisible by 2, the second by 3 (because both $100!$ and 3 are divisible by 3), the third by 4, and so on. The last is divisible by 100. There we have 99 consecutive integers, rather large ones it's true, that are all composite. If we had started with $1,000,000!$ we would have $999,999$ integers in a row that are composite. Recently a prime gap of length 337446 has been discovered between a pair of 7996-digit integers. Those are considerably smaller than $337447!$, which has more than $1,500,000$ digits."

"Is $100! + 101$ prime?" I asked.

"No one knows, nor is anyone likely to find out any time soon." Richard said. "But if you're going to bet, bet against it. The famous Prime Number Theorem says, roughly, that the proportion of integers around n that are prime is $1/\ln n$. So, around $100! + 101$ fewer than 1 in every 350 integers is prime. $100! + 101$ may be one of them, but it isn't likely."

"How can we tell if an integer is prime?" I asked.

"That is an important question," Richard answered, "that people are working on all the time. There exist powerful methods, so powerful that we—we and our computers, that is—can determine the primality or lack of it of 100-digit integers very quickly. Of course the question of primality was in a sense settled long ago by Wilson's theorem that states that $(n-1)! + 1$ is divisible by n if and only if n is a prime. A necessary and sufficient condition! For example, if $n = 5$, $(5-1)! + 1 = 25$. That is divisible by 5, so 5 is a prime. For $n = 6$, $(6-1)! + 1 = 121$. That is not divisible by 6, so 6 is not prime. Thus, to find out if n is prime, just calculate $(n-1)! + 1$ and see if it's a multiple of n. The difficulty is that the calculation isn't practical for large integers. A really pure mathematician—the Platonic ideal of purity—would say that the problem of primality has been completely solved so we don't have to bother with it any more. But I think that most actual mathematicians are interested in the progress being made in primality testing and factoring. It has generated some very nice mathematics. Or, as I should say, it has caused some very nice mathematics to be discovered. As Wittgenstein once said—"

Richard was showing signs of going into his philosophical mode. This would provide no material for my column, so I interrupted.

"Richard, can you give me anything interesting about 12? It's one less than a baker's dozen and one more than the smallest prime all of whose digits are 1—what else would my readers find interesting?"

"Ah, Owen, 12 has dominated in the affairs of the human race for centuries," Richard said. "There were twelve tribes of Israel. There were twelve apostles. There are twelve months in one year. There are twelve signs of the zodiac. There are twelve persons (usually) in a jury. Grocers are merchants who deal with quantities by the gross—twelve twelves. There are twelve letters in the Hawaiian alphabet, twelve faces on the dodecahedron, twelve face cards in a deck of cards, twelve inches in one foot, and twelve items in one dozen. In the days of pre-decimal currency in Ireland and the U.K. (before 1971) there were twelve pennies in one shilling. To the present day only twelve humans have walked on the moon."

"What about the mathematical properties of the number 12?" I asked.

"The number 12 is divisible by the sum of its digits and by the product of its digits," Richard said. "As is twice twelve. Your readers might like to try to find other examples. 12 squared is 144, and the reversal of 12 squared is 441, the reversal of 144. The product of the proper divisors of 12 is 12^2. The smallest number that is less than the sum of its factors (excluding itself) is 12."

Richard was not being as brilliant as was his wont. I decided to change the topic, by adding 5 to 12.

"Very good. Now how about seventeen?"

"It occurs in many places in history. The last mission to the moon was on Apollo 17 in 1972. Note that $1972 = 17 \cdot 116$. The period of revolution of

Callisto, the satellite of Jupiter discovered in the 17th century by Galileo, is just under seventeen days. The French mathematician, Jean-le-Rond d'Alembert, was born in 1717. The element selenium, whose atomic number is twice 17, was discovered by Jons Berzelius in 1817, and of course $1 + 8 + 1 + 7 = 17$. On November 17, 1972 General Peron went back to Argentina after a seventeen-year exile. The Demilitarized Zone in Vietnam was created along the seventeenth parallel in April 1954. The Lebanon war lasted seventeen years (1975–1992). The IRA ceasefire that started in 1994 lasted for seventeen full months. The atrocity that killed the most people in any one day in the recent Irish troubles occurred on May 17th, 1974. Seventeen-year locusts have evolved a prime number for their periodicity.

"The integer 17 has some unusual properties. It is $1^2 + 4^2$, $2^3 + 3^2$, and $3^4 - 4^3$. It lies just between a square, 16, and twice a square, 18. The square of 17 can be expressed as the sum of 1, 2, 3, 4, 5, 6, 7, or 8 distinct squares. The first four digits of $2^{17} \cdot 2^{17}$ are 1717. The primes up to 17 can be partitioned into two sets such that the difference of their product is unity: $(5 \cdot 11 \cdot 13) - (2 \cdot 3 \cdot 7 \cdot 17) = 1$.

"As a change of pace from all of that true information, your readers might like something false. Perhaps not all of them have seen this proof that $1 = 2$.

Let

$$a = b.$$

Multiply by b:

$$ab = b^2.$$

Subtract a^2:

$$ab - a^2 = b^2 - a^2.$$

Factor:

$$a(b - a) = (b + a)(b - a).$$

Cancel the common factor:

$$a = b + a.$$

Since $b = a$ we get

$$a = a + a = 2a.$$

Divide by a:

$$1 = 2.$$

Ask your readers if they can spot the error in the reasoning. There had better be one, since if $1 = 2$ it follows that all integers are equal and hence that all fractions $a/b = 1$. From that it's an easy step to showing that all square roots of integers are equal, because $\sqrt{n} = \sqrt{\frac{n}{1}} = \sqrt{1} = 1$. I'd leave the implications about transcendental numbers like e and π to your readers, but they wouldn't have time to get to them: the universe would already have collapsed. After all, if $\sqrt{2} = 1$ then the diagonal of a square with side 1 has the same length as the side, which means that squares become straight lines. So do cubes. Everything has collapsed inside black holes—I wonder what mathematics is like there. In any event, tell your readers that they need to hurry to find the error before it's too late."

"I will, Richard," I said. "While we're waiting for the collapse, do you have any good curiosities with factorials?"

"Need you ask?" said Richard. "Only four integers are the sum of the factorials of their digits: 1, 2, 145 ($1! + 4! + 5! = 1 + 24 + 120$), and 40585 ($4! + 0! + 5! + 8! + 5! = 24 + 1 + 120 + 40320 + 120$). $0! = 1$, not from the definition $n! = n(n-1)(n-2)\ldots 1$, but because it's defined that way so as to make some factorial formulas give the right result when 0 is put into them. By the way, $10! = 6! \cdot 7!$. Here are some equations that I think are pretty, with the nine digits in ascending and descending order. With one factorial,

$$1 + 23 + 4! = 56 - 7 + 8 - 9$$
$$98 - 7 - 65 = 4! + 3 - 2 + 1.$$

With two factorials,

$$1 \cdot 2 + 3! + 4! = 56 - 7 - 8 - 9$$
$$98 - 76 = 5!/4 - 3! - 2 \cdot 1.$$

Ask your readers if they can find two similar equations, with each containing three factorials. It's even possible to have the factorials appear in successive terms. The rule is that, besides the factorial sign, only plus, minus, times, and divides are allowed."

"I suppose asking for four factorials would be too much," I said.

"It might be," said Richard, as usual giving nothing away. "Factorials arise naturally in finite differences, a subject not studied as much as it used to be.

Computers have killed it off, which is too bad because it contained (and still contains, for that matter, but mostly in dusty and unread library books) many nice things. Take the sequence of cubes, and make a difference table:

n	n^3	Δn^3	$\Delta^2 n^3$	$\Delta^3 n^3$
1	1			
		7		
2	8		12	
		19		6
3	27		18	
		37		6
4	64		24	
		61		6
5	125		30	
		91		
6	216			

The third column of differences is 3!. Here's a similar table for fourth powers, with the difference columns moved upward to save space:

n	n^4	Δn^4	$\Delta^2 n^4$	$\Delta^3 n^4$	$\Delta^4 n^4$
1	1	15	50	60	24
2	16	65	110	84	24
3	81	175	194	108	24
4	256	369	302	132	
5	625	671	434		
6	1296	1105			
7	2401				

The last column is constant at 4!, and so it goes in general, $\Delta^k n^k = k!$. I remember when I was a child, putting numbers down more or less at random and making difference tables, watching the numbers go up and down. Ah, the golden and innocent days of childhood, when the entire world was new, and difference tables and other delights lay all around, waiting to be explored."

When I was a child I spent more time with balls than with difference tables, but it wouldn't do to suggest to Richard that he was in any way odd.

"There is a very nice formula, Stirling's formula," the Professor said, "that gives the approximate value of a factorial. It involves both e (2.7182818...) and π (3.14159...), which shows how ubiquitous those numbers are. It is

$$n! \approx n^n e^{-n} \sqrt{2\pi n}.$$

It's an approximation and is not useful for finding numerical values. When $n = 5$, it gives

$$5^5 e^{-5} \sqrt{2\pi \cdot 5} = (3125)(.0067369)(5.605) = 118.01\ldots,$$

which is not terribly close. The percentage error decreases as n increases, though. It can be useful when you want to know, roughly, the size of something. For example, how big is $\binom{2n}{n}$? Very big, of course, when n is big, but how is it behaving? Like n^2, or e^n, or what? Stirling's formula lets us get a handle on it:

$$\binom{2n}{n} = \frac{(2n)!}{n!n!} \approx \frac{(2n)^{2n} e^{-2n} \sqrt{2\pi(2n)}}{\left(n^n e^{-n} \sqrt{2\pi n}\right)^2} = \frac{2^{2n+1} n^{2n} e^{-2n} \sqrt{\pi n}}{n^{2n} e^{-2n} (2\pi n)}$$

$$= \frac{2^{2n}}{\sqrt{\pi n}} = \frac{4^n}{\sqrt{\pi n}}.$$

So, it increases a good deal faster than e^n but not quite as fast as 4^n."

"That might be a little too advanced for my readers," I said.

"Give it to them anyway, Owen. People should be stretched now and then. If they're not stretched, they'll stay forever small."

"That's true. In any event, it's interesting how factorials, which involve only integers, should be related to e and π. Those are numbers that come from logarithms and circles, not from arithmetic."

"Indeed so," said Richard. "All mathematics is one."

It was a perfect curtain line, but he didn't want to get off the stage.

"Do you know what a self-descriptive integer is, Owen?" he asked. "I assume not, so I'll tell you. It is one that has the following very rare property. Digit n is exactly the number of ns in the integer, where we number the digits starting at 0."

"What?" I said.

"An example would be helpful, I suppose. Consider 6210001000," he said. "With its digits numbered, we have

Digit	0	1	2	3	4	5	6	7	8	9
Number of appearances	6	2	1	0	0	0	1	0	0	0

Do you see? 6210001000 has six 0s, two 1s, one 2, no 3s, 4s, or 5s, one 6, and no 7s, 8s, or 9s. It describes itself."

"So it does," I said. "Now you're going to tell me to ask my readers to find other self-descriptive numbers."

"No," he said, "that is the only ten-digit self-descriptive number. It's possible to consider variations, of course—fewer than ten digits, numbering the digits from 1 instead of zero, looking at bases other than 10—but I have nothing up any of my sleeves about them."

"It's almost 11," I said, "and the bar will be closing soon. We'll have to stop."

"Yes, I haven't heard the swish of darts for quite some time," Richard said. "Did you know that in New Orleans the bars traditionally never had to close? Ireland still has some patches of Puritanism. I'll save what I haven't told you until next time."

Solutions

1. The three darts questions are answered as follows.

    (a) The lowest score that cannot be scored with one dart is 23.

    (b) The lowest score that one may be left with in darts where the game cannot be finished in two darts is 99.

    (c) The lowest score that one may be left with in darts where the game cannot be finished in three darts is 159.

2. The mark is doubling each of his scores, so he is always subtracting an even number from an odd number and so will always be left with an odd number. Thus he will never be able to reach zero with a double.

3. Assume that the score of 100 can be scored with three triples. Call the scores a, b, and c. Then

$$3a + 3b + 3c = 100$$

or

$$a + b + c = 33.3333\ldots.$$

But a, b, and c are integers. The sum of three integers must equal an integer, so this is impossible. Therefore our original assumption must be false. In other words, it is not possible for three trebles to equal 100.

4. The 26 ways of scoring 26 to go out with two darts are
    (single, double): (2, 12), (4, 11), (6, 10), (8, 9), (10, 8), (12, 7), (14, 6), (16, 5), (18, 4), (20, 3)
    (double, double): (1, 12), (2, 11), (3, 10), (4, 9) (5, 8), (6, 7), (7, 6), (8, 5), (9, 4), (10, 3), (11, 2), (12, 1)
    (triple, double): (2, 10), (4, 7), (6, 4), (8, 1).

5. The chance in the cards hustle can be calculated as follows. If there is no pair in the five cards dealt, all five cards have different denominations. We first determine the chance of this happening. The first card dealt from the 52-card deck can be any one of 52. The second card dealt must not have the same denomination as the first if there is to be no matching pair. Therefore the second card can only be one of 48. The third card, for the same reason, can only be one of 44, the fourth one of 40, and the fifth one of 36. The order of the cards is irrelevant. Therefore we divide the product of these numbers by the number of different orders in which they could appear, 5! or 120, to get $(52 \cdot 48 \cdot 44 \cdot 40 \cdot 36)/120 = 1,317,888$ ways that five cards can be dealt from a deck of 52 cards so that there are no matching pairs. There is a total of

$$\binom{52}{5} = \frac{52 \cdot 51 \cdot 50 \cdot 49 \cdot 48}{5!} = 2,598,960$$

poker hands in a deck. So, 1,317,888/2,598,960, or 50.7% per cent of 5-card hands dealt in such a fashion do not contain a pair and 1,281,072, or 49.3% do.

On the other hand, the total number of ways six cards can be dealt from a 52-card deck is

$$\binom{52}{6} = \frac{52 \cdot 51 \cdot 50 \cdot 49 \cdot 48 \cdot 47}{6!} = 20,358,520$$

and the number of ways of dealing six cards from a deck so that no hand contains a pair is $(52 \cdot 48 \cdot 44 \cdot 40 \cdot 36 \cdot 32)/720 = 7,028,736$. Therefore, there are 13,329,784 ways that six cards can be chosen from 52 where each hand does contain a pair. This means that when six cards are dealt from a 52-card deck, the chance is 65.5% there will be a pair in the hand. The odds are nearly 2 to 1 in favour of the hustler.

6. The puzzle concerning the sheet of paper has a very surprising answer. The height of the stack of paper would be more that 177 million miles! The first cut produces two pieces, the second cut produces four pieces, the third cut eight, the fourth cut 16, and so on. The number of pieces after the nth cut is 2^n. Therefore, after 50 cuts the number of pieces is 2^{50}. Now 2^{50} is a number containing 16 digits. Every hundred of these pieces equals one inch. So divide 2^{50} by 100 to obtain the height of the stack in inches. Then divide the answer by 12 to convert to feet. Then divide that by 5280 to convert the height of the stack to miles. The answer is that the stack of paper is 177,698,848.9 miles high.

Many people find the answer to this puzzle surprising because they are not familiar with how quickly numbers increase when repeatedly doubled. This doubling principle crops up in many old teasers.

An old chestnut tells of a farmer who brought his horse to the blacksmith to be shod. The farmer is told by the blacksmith that the first nail will cost a cent, the second nail two cents, the third four cents, and so on, each nail being double the cost of the previous one. If a total of twenty-eight nails are required to shoe the horse, how much will this transaction cost the farmer? The cost of shoeing the horse in cents is

$$1 + 2 + 2^2 + \cdots + 2^{27} + 2^{28} = 2^{29} - 1,$$

or $5,368,709.11.

7. The fallacy in the algebra problem arises when we attempt to divide the equation by $b - a$. We are told that $a = b$, so $b - a$ equals zero. We are then told to divide by $b - a$, but we cannot divide by zero. If we could, we could conclude, as in this fallacy, that from $a \cdot 0 = (a + b) \cdot 0$ (which is true) it follows that $a = a + b$ (which is false, unless $b = 0$).

8. Here are the two equations the professor gave me.

$$1! + 2! + 3! + 4 = -56 + 78 - 9$$
$$98 - 7!/6! = 5! + 4 - 32 - 1.$$

References for further reading

Martin Gardner, *Mathematical Magic Show*. Penguin Books, 1985, Chapter 4 "Factorial Oddities".

E. A. Maxwell, *Fallacies in Mathematics*. Cambridge University Press, 1959.

CHAPTER 12

The professor gives some number patterns

"Owen, wouldn't you like to go on an excursion?" It was Richard on the phone.

"Where to?" I asked

"Not far, just to Mallow," he said. Mallow is only a bit more than twenty miles north of Cork.

"Why Mallow?" I asked.

"I could say to see the marsh there, but that wouldn't be the real reason," he answered.

"What marsh?" I asked. Telephones rattle me—I like to see the person I'm talking to—so I didn't get the joke until after I had hung up. It no doubt gave Richard some pleasure that I didn't see it instantly.

"Never mind," Richard said. "Mallow is the birthplace of Robert Murphy and I'd like to see if there are any traces of him left there."

"Who's Robert Murphy?" I asked. I was not at my best. All I could do was ask obvious questions.

"Ah, Owen, shame for not knowing about one of Ireland's best mathematical minds," Richard said. I knew that my lack would shortly be remedied.

"Murphy was born in Mallow in 1806, the son of a shoemaker," the professor said. "His unusual mathematical talent was noticed by some of the worthies of the town, who sent him to Cambridge in 1825. He was third in the university mathematical competition in 1829, and he was asked to stay on as a fellow. Unfortunately, he seems to have been one of those wild and irresponsible Irishmen. 'A course of dissipation led him into debt' a 1911 reference says. I wish those old reference works were more specific and gave the juicy details. Leaving them out can make you think worse things than what actually happened. Cambridge threw him out in 1832, he retreated to Ireland, but returned to England in 1836. He did a few useful things mathematically, but his dissipation had either taken its toll or (more likely) he continued as before as much as his finances allowed, and he died a very early death in 1843. Talent wasted!"

"And an Irish talent having to leave the country in order to be used," I said. "The old story. Let's go to Mallow. I'll come by to pick you up."

We didn't find any statues of Robert Murphy in Mallow, but Richard didn't seem to mind.

"Talent, even genius, often goes unrecognized," he said. "I doubt that I'll rate more than a plaque or two after I'm gone. But now it's time for dinner. I know just the place: Keppler's Restaurant. The extra p probably means that the proprietor is not a descendant of the great Johannes, but we should support mathematics in every way we can, even indirectly. Besides, it's a good place for me to give you some of the things I've had in my head recently—it has the Noggin Bar."

"Richard," I said, "you go too far. The name undoubtedly refers to the liquid measure. A quarter of a pint, I think."

"You can't be sure, Owen," he replied. "Kepler, the head—they go together, you know."

After we settled in and Richard had observed that the restaurant was missing a bet by not listing appetizers under "Keppler's First Law," and so on (it would have gone over the heads of the patrons, I thought), he proceeded to give me some interesting information.

"You know I like to give my readers unusual proposition bets now and again," I said. Any suggestions?"

"Try this one for size," said Richard. "Give someone, let's say your twin brother, Michael, two dice to toss. Before he tosses ask him to name two numbers from 1 to 6. Tell him that you are prepared to bet even money that at least one of the two numbers he chose will turn up when the dice are tossed. The mark will reason that the odds favor him. Ask your readers to calculate the true and surprising odds."

"I will," I said. "Today is the 19th, so I was wondering if you know of any interesting information about 19. Try not to talk nineteen to the dozen when telling me."

"I always speak deliberately, Owen." (Except when excited, I thought.) "Of course 19 crops up here and there in the everyday affairs of humans. Nineteen lamps, for instance, illuminate the Statue of Liberty's torch in New York harbor. Nineteen years of 365.2422 days are very close to 235 lunar months of 29.5306 days. The radioactive gas oxygen 19 has a half-life of approximately 27 or $(19 - 1 + 9)$ seconds. Bela Bartok completed his Opus 19 in 1919, when he was thirty-eight—or twice 19—years old. Edwin Aldrin, the second man to walk on the moon, walked on the lunar surface nineteen minutes after Neil Armstrong first set foot on the moon. Usama Bin Laden has nineteen brothers. Nineteen hijackers carried out the 9/11 atrocities."

"I presume you will give me some mathematical properties of nineteen now," I said.

"Only if you want me to," said the professor.

"Shoot," I said.

"Let's see. The number formed with nineteen ones, 1111111111111111111, is a prime. The first two digits of 19^{19} are 1 and 9. $19 = (1 + 9) + (1 \cdot 9)$. The smallest number that can be expressed as the sum of two cubes in two ways is 1729, as everyone who has heard the anecdote about Hardy and Ramanujan knows. But even Ramanujan may not have noticed that 1729 is 19 multiplied by its reversal, 91. The number 19 is the only prime equal to the difference of two prime cubes: $3^3 - 2^3$. Also, 19 is the $(-1 + 9)$th prime. The digits of 19 and of its square, 361, sum to 10. All positive integers are the sum of at most 19 fourth powers and $19 = 4! - 3! + 2! - 1!$ Also, $19 = 8 + 2 + 8 + 1$ and its reverse, 91, is $\sqrt{8281}$."

"And," I said, in order to not be left out, "a cribbage hand can score any number of points from 0 to 24 except for 19."

"Quite right," Richard said. "*Cribbage*, by the way, is one of those words whose origins are uncertain. If your mind runs that way, you might think that this implies that the game was brought here, complete with name, by aliens. *Aliens*, by the way, is one of those words that truncates nicely: *aliens-snail-nail-ail-ai-a*. An *ai* is a three-toed sloth. Speaking of words (as well as speaking *in* words, which is not the same as speaking in tongues), ask your readers to find the longest palindromic word in fairly common use in the English language. Do you recall Lewis Carroll's word ladders? To change a *fool* into a *sage* one letter at a time, go *fool-pool-poll-pall-pale-sale-sage*. Ask your readers to do something similar and change *ape* into *man* with five intermediate steps."

"I will," I said.

"My mind is racing," Richard said. "For some reason, my namesake, Richard Feynman comes to mind."

"You admire him," I said.

"I do indeed," Richard said. "Feynman was brilliant. I like his sense of playfulness towards discovering the secrets of nature. However, I do disagree with Feynman's view of the scientific enterprise. He was inclined to believe that humans would eventually reach the stage where all the major discoveries about science would become known, and from that time onwards it would merely be a question of filling in the details, so to speak. Of course, this is impossible in mathematics, where every question answered generates ten new ones, but in physics, I think that we can approach ultimate truth only asymptotically, without ever actually attaining it. It is my belief that we will *never* know the true nature of things."

"I think," I said, "that I would be inclined to agree with you there."

"Incidentally," said Richard, "it was pointed out to Feynman when he was young that $\cos 20° \cos 40° \cos 80° = \frac{1}{8}$, exactly. He remembered that curiosity for the rest of his life. He was a marvelous character."

Richard was slowing back to normal.

"Mathematics," said the professor, "is a most astonishing and amazing subject. Human beings, in my opinion, do not invent mathematics. We *discover* it. Mathematical truth, in my view, has always existed. It is eternal. Einstein said it aptly when he said 'Politics is for the moment, an equation is forever.' I have always found it amazing that the deepest and most profound mathematical equations discovered are beautiful. If something is beautiful, it does not necessarily mean it is true. But if a mathematical theorem is true, then you can bet your bottom dollar that it is beautiful. It is very difficult to define mathematical beauty, especially since a distressingly large fraction of the population lacks the mathematical aesthetic sense. But most, if not all, top rate mathematicians in the world will tell you that mathematical truth is beautiful."

"You appear to have a great love for mathematics, Richard."

"How can any thinking being not? I have always found it astonishing, for instance, that mathematics is the language of nature. It is a marvelous mystery that nature seems to obey mathematical equations. For example, take Kepler's Third Law in astronomy. Consider the distance of the planets in the solar system from the sun in relation to the Earth's distance. Jupiter for example is approximately 5.2 times further from the sun than the Earth. The time Jupiter takes to orbit the sun is the square root of 5.2 cubed, about 11.85 years. Saturn's average distance from the sun is 886.7 million miles, so it is about 9.534 times further from the sun than the Earth. The time Saturn takes to orbit the sun is $\left(\sqrt{9.534}\right)^3$, 29.43 years. All the planets in the solar system follow this simple mathematical rule. Kepler saw the pattern."

"How did he do that?" I asked.

"How does anyone see a pattern?" Richard answered. "You *see* it. Or you don't. Most of us usually don't. Whoever first noticed this pattern

$$1 + 2 = 3$$
$$4 + 5 + 6 = 7 + 8$$
$$9 + 10 + 11 + 12 = 13 + 14 + 15$$
$$16 + 17 + 18 + 19 + 20 = 21 + 22 + 23 + 24$$
$$\cdots$$

made a fine discovery. It is a lovely number pattern that every child in the world who is taught mathematics should be made aware of. Adults would do well to know about it too. It illustrates the simple harmony and sheer elegance found in numbers. It's easy to recall. Each row begins with the square of the number of its row, the left-hand side of row n has $n + 1$ terms, and the right-hand has side has n. The sum in the first row is 3, in the second row is 15, the third row, 42, and the fourth, 90. Ask your readers if they can find a simple and pretty expression that gives the sum of the numbers in any row.

"There is another beautiful pattern," said Richard, "involving the sums of squares. It illustrates the intimate connection between odd numbers, square numbers and triangular numbers. But first take a look at the sequence of triangular numbers:

$$1 = 1$$
$$1 + 2 = 3$$
$$1 + 2 + 3 = 6$$
$$1 + 2 + 3 + 4 = 10$$
$$\cdots$$

The triangular numbers are 1, 3, 6, 10, 15, 21, 28, 36, ... and so on to infinity. Now look at this:

$$3^2 + 4^2 = 5^2$$
$$10^2 + 11^2 + 12^2 = 13^2 + 14^2$$
$$21^2 + 22^2 + 23^2 + 24^2 = 25^2 + 26^2 + 27^2$$
$$36^2 + 37^2 + 38^2 + 39^2 + 40^2 = 41^2 + 42^2 + 43^2 + 44^2$$
$$\cdots$$

"The first row gives the smallest Pythagorean triple. The other sums of squares are not so well known. There are some interesting curiosities in this pattern. First, note that the first numbers on the rows are every other triangular number 3, 10, 21, 36, and so on. Second, note that the number on the extreme right of row n equals four times the nth triangular number, plus the number of the corresponding row. For example, the number on the extreme right of row 1 equals four times the first triangular number, plus 1. The number on the extreme right of row 2 equals four times the second triangular number, plus 2. The number on the extreme right of row 3 equals four times the third triangular number, plus 3. And so on. Consider also each of the numbers immediately to the left and right of the equation sign. They are members of Pythagorean triples:

$$3^2 + 4^2 = 5^2, \quad 5^2 + 12^2 = 13^2, \quad 7^2 + 24^2 = 25^2, \quad 9^2 + 40^2 = 41^2,$$

and so on. The first numbers are the squares of the successive odd numbers, the second are the squares of four times the successive triangular numbers, and the third is one more than the second.

"Here is another beautiful number pattern. It shows an intimate relationship between the odd numbers and the cubes.

$$1 = 1^3,$$
$$3 + 5 = 2^3,$$
$$7 + 9 + 11 = 3^3,$$
$$13 + 15 + 17 + 19 = 4^3,$$

and so on. Ask your readers if they can find a simple expression that gives the numbers to the left of the equals signs, $1, 5, 11, 19, \ldots$."

"I will," I said. "You know, in all of our discussions, we haven't singled out 1 for special attention. Has that been on purpose?"

"No indeed," said the professor. "It takes time to get around to all the integers. We're probably going to have to ignore some of them, but not 1. It's the only positive integer that is neither prime nor composite. Some representations of 1 with reciprocals are

$$\frac{1}{2} + \frac{1}{3} + \frac{1}{6}, \quad \frac{1}{2} + \frac{1}{4} + \frac{1}{6} + \frac{1}{12}, \quad \frac{1}{2} + \frac{1}{4} + \frac{1}{8} + \frac{1}{16} + \ldots, \quad \text{and}$$

$$\frac{1}{1 \cdot 2} + \frac{1}{2 \cdot 3} + \frac{1}{3 \cdot 4} + \frac{1}{4 \cdot 5} + \ldots.$$

Take any four consecutive terms of the Fibonacci sequence $1, 1, 2, 3, 5, 8, 13, 21, \ldots$. The product of the two outside numbers always differs from the product of the two inside numbers by 1, as in $3, 5, 8, 13$ we have $3 \cdot 13$, one away from $5 \cdot 8$. The larger n is, the closer the nth root of n gets to 1. The probability that something happens is 1 minus the probability that it doesn't happen. We would have a hard time getting along without 1."

"I suppose that if 1 were to go missing, we'd notice it sooner than if something like 8506687126 disappeared," I said.

"Perhaps," Richard said, "but the consequences would be equally grave no matter which integer went away, even if it were one that you had never heard of."

"How's that?" I asked.

"For one thing, proofs by mathematical induction would be more difficult. If 17 were to disappear, we'd have to do them in two parts, starting with $n = 1$ and then again with $n = 18$. For another, the sum of two integers would no longer always be an integer. The Fundamental Theorem of Algebra, that every polynomial equation has a root, would go by the boards. There would be chaos. Let us give thanks that we have with us all the integers, from 1 on up, and that none will ever leave us. By the way, 1 is a big help in finding the number of factors of an integer. Take 24 for example. Its prime-power factorization is $24 = 2^3 \cdot 3$. Add 1 to each exponent and multiply the results: $4 \cdot 2 = 8$, the number of factors of 24."

"Is that right?" I asked. "1, 2, 3, 4, 6, 8, 12, and 24—yes, there are eight of them. So since $60 = 2^2 \cdot 3 \cdot 5$ it's going to have $3 \cdot 2 \cdot 2$ factors?"

"Exactly," Richard said, "and you don't have to find them to know how many there are. The largest fraction with a numerator of 1 is, of course, 1/2. There are just 12 (note the two digits) nine-digit fractions with value 1/2:

$$\frac{6729}{13458}, \frac{6792}{13584}, \frac{6927}{13854}, \frac{7269}{14538}, \frac{7293}{14586}, \frac{7329}{14658},$$

$$\frac{7692}{15384}, \frac{7923}{15846}, \frac{7932}{15864}, \frac{9267}{18534}, \frac{9273}{18546}, \frac{9327}{18654}.$$"

"Richard," I said, "you are a mine of information."

"Well, Owen," said the professor, "numbers are my friends."

"And good ones to have," I said. "They don't let you down. Nor do they run away. Incidentally Richard, you know that the centenary of Bloomsday, June 16, 1904 occurred not long ago. Have you any comments on James Joyce or his famous book, *Ulysses,* that you wish to make?"

"I certainly do," Richard said. "I have always admired Irish writers. You've probably forgotten about, or never even heard of, Jim Prendergast, the Irish communist. He was made to stand trial in 1969 on a frivolous charge—it was harassment, really—of writing a threatening letter. Anticipating a possible defense, the prosecution suggested that he, an ignorant Irishman, might not be aware of the meaning of what he wrote. He magnificently said, 'Your Honor, like Oliver Goldsmith, Dean Swift, Edmund Burke, Oscar Wilde, William Butler Yeats, George Bernard Shaw, John Millington Synge, Sean O'Casey,

James Joyce, Brendan Behan, and Samuel Beckett, I occasionally have difficulties with the English language.' Joyce was a literary genius, and *Ulysses* is his masterpiece. He was a lover of all classes of puzzles, Owen, both mathematical and linguistic. He was particularly fond of palindromes. I often wonder if Joyce was aware of how 2, or the palindrome 22, seemed to crop up quite a lot in his life. Did you know, for instance, that Joyce was born on 2/2 1882? He was just 22 years old in 1904 when he met his wife, Nora Barnacle. He was 40 or $22 + 22 - 2 - 2$ years old when *Ulysses* was first published in the 22nd year of the 20th ($22 - 2$) century. The first two copies of Ulysses were given to Joyce just before 2/2 1922, in time to celebrate Joyce's fortieth birthday. Joyce's full name was *James Augustine Aloysius Joyce*, with palindromic initials. Using the code $a = 1, b = 2, c = 3$, and so on, the sum of the value of the initials *JAAJ* is 22."

"Numerical palindromes with an even number of digits are always divisble by 11," I said. "I didn't know that extended to words."

"I trust you jest, Owen," Richard said. "The value of *boob* is 34, not a multiple of 11. The whole of *Ulysses* is based on the events of one day, June 16, 1904, in Dublin. That day is now universally known as Bloomsday. Joyce apparently chose June 16 because he and Nora Barnacle commenced their relationship on that date. June 16, 1904 was the 168th day of that year. I wonder if Joyce was aware that 168 equals $(2 + 2 + 2 + 2) \cdot 22 - (2 + 2 + 2 + 2)$. Was Joyce aware, I wonder, that June 16 may be written as 6/16 or 616, which is a palindrome? The sum of 6 and 16 is 22 and $616 = 28 \cdot 22$. Joyce's first and last names begin with the 10th letter of the alphabet,

$$10 = 1 + 2 + 3 + 4, \quad \text{and} \quad 1!^2 \cdot 2!^2 + 3!^2 + 4!^2 = 616,$$

the date of Bloomsday. When Joyce died in Zurich on January 13, 1941 he was 58 years old. The digits of 58 sum to 13, his day of death, whose digits sum to 4, or $2 + 2$. His mother also died on the 13th of the month. Joyce died in 1941, and $41 - 19 = 22$.

"Joyce's date of birth (ignoring the year) is 2/2, or 22. Its square, $22^2 = 484$, is a palindrome. His date of death (ignoring the year) is 1/13, or 113. Its square, $113^2 = 12769$, isn't a palindrome, but its reverse, 96721, has for its square root 311, which is the reverse of 113. Here is a little number curiosity concerning Joyce that your readers may enjoy. The two words *James Joyce* contain 10 letters. Multiply 10 by 22, obtaining 220. Joyce was born in 1882 and died in 1941. Multiply 1882 by 1941, obtaining 3652962. Divide this by 220. The answer is 16604.3727. . . . Ignoring the decimal part we have 16,604 or 16/6/04, the date of Bloomsday as it is usually written in Europe."

Richard looked at his watch. "Time passes quickly," he said. "Incidentally, do you consider it strange that we should call the third hand on a watch the second hand? Here is one more little puzzle that may interest your readers. Suppose one million people were tossing coins at the rate of ten coins per minute, forty hours per week, fifty-two weeks per year. How often could we expect to see a coin falling heads fifty times in a row?"

"That's a nice puzzle, Richard. I will give that to my readers to solve."

"Here's another one, appropriate for the year 2005. 2005 is the sum of two consecutive integers, 1002 and 1003, and also the sum of five consecutive integers, 399 to 403. Ask your readers if they can find ten consecutive integers that sum to 2005. I'll give a little hint—the first integer lies between 100 and 200."

"I hope they won't try all 100 possibilities," I said.

"They will be more clever than that," said Richard. "Well, Owen, this has been a pleasant excursion even though we found no lost manuscripts of Robert Murphy. Now it's time to be getting back towards Cork, I think."

On the way back, Richard closed his eyes, I'm sure to concentrate on something new he could give me at our next meeting. It will be a sad day when he leaves Ireland.

Solutions

1. Two dice are about to be tossed. The mark names two numbers. You bet him even money that at least one of the numbers will turn face up on the tossed dice. If the mark accepts the bet at even money he will be making an unfavorable bet. The chance that any one of the two numbers will not turn face up on the throw of one die is clearly 4/6, or 2/3. The chance that any one of the two numbers will not turn face up on the throw of two dice is thus $(2/3)^2$, or 4/9. Therefore the chance that one of the two numbers will turn face up on the toss of two dice is 5/9, or 55.5. . . %, so the bet is clearly to your advantage.

2. We were asked to change the word *ape* to the word *man* in five steps, altering only one letter at each step. The problem can be solved with

   ape, apt, opt, oat, mat, man.

3. The longest English single word palindrome in more or less common use is *redivider*.

4. The sum of the numbers in row 1 is 3, in row 2 is 15, in row 3 is 42, in row 4 is 90, and so on. The reader was asked to find a pretty expression that gives

the sum of the numbers in any row. Beauty of course is in the eye of the beholder, but it would be difficult to argue that the following is not pretty:

$$3 = 1 + 1^2 + 1^3,$$
$$15 = 1 + 2 + 2^2 + 2^3,$$
$$42 = 1 + 2 + 3 + 3^2 + 3^3,$$
$$90 = 1 + 2 + 3 + 4 + 4^2 + 4^3,$$

and so on. The sum of the numbers in row n is the sum of the numbers from 1 to n, plus n^2, plus n^3, or, as it could be written,

$$\frac{n(n+1)}{2} + n^2 + n^3 = \frac{n(n+1)(2n+1)}{2}.$$

5. The expression that the professor spoke of concerning the numbers on the immediate left of the equation sign in the number pattern relating to the odd numbers and cubes is as follows:

$$1 = 1^2 + 0,$$
$$5 = 2^2 + 1,$$
$$11 = 3^2 + 2,$$
$$19 = 4^2 + 3,$$

and so on, or $n^2 + n - 1$, $n = 1, 2, 3, \ldots$.

6. First consider a coin tossed once. There is one chance in two that a head will turn up. With two tosses there is one chance in 2^2 of getting two heads in a row. With fifty coins, there is just one chance in 2^{50} of getting fifty heads in a row.

   In other words, if we were to toss one coin 2^{50} times, we could expect 50 heads in a row to occur just once. One million persons tossing a coin at the rate of ten per minute, forty hours per week, fifty-two weeks per year, is the equivalent of $1000000 \cdot 10 \cdot 60 \cdot 40 \cdot 52 = 1248000000000$ tosses. We divide 2^{50} by 1248000000000. Our answer is $902.1633\ldots$. This tells us that if the above experiment could be carried out we could expect fifty heads to fall in a row in fifty tosses once every 902 years. This gives us some idea of how unlikely it is that fifty tosses will produce fifty heads in a row.

7. The answer to how to express 2005 as a sum of ten consecutive integers is

$$2005 = 196 + 197 + 198 + 199 + 200 + 201 + 202 + 203 + 204 + 205.$$

To find this without using trial and error, note that the sum of n consecutive integers starting with a is

$$a + (a+1) + (a+2) + \cdots + (a + (n-1))$$

$$= na + (1 + 2 + \cdots + (n-1)) = na + \frac{(n-1)n}{2} = \frac{n(2a+n-1)}{2}.$$

So, putting $n = 10$ in $2005 = \frac{n(2a+n-1)}{2}$ gives $401 = 2a + 10 - 1$, so $a = 196$. Readers may find it amusing to write 2006 as a sum of 17 consecutive integers.

Reference for further reading

David Wells, *The Penguin Dictionary of Curious and Interesting Numbers*. Penguin Books, 1986.

CHAPTER **13**

The King James Bible and some currency curiosities

"Michelle has worn me down," Richard said. "We have to go to Blarney Castle. 'It's less than an hour from Cobh,' she said, among many other things. 'How can we be this close and not go?' "

"She has a point, you know," I said. "Surely you want to kiss the Blarney Stone."

"Of course I don't," said the professor. "Wait in line for what might be an hour to climb up a cramped spiral staircase five stories high with only a rope to hold onto? No thank you—I've read reports of what the site is like. And then bend over backward with someone holding on to your legs to keep you from falling to your death, just so you can put your lips on a stone that has had millions of other people slobber over it? How absurd!"

"Millions of people can't all be wrong, can they?" I asked.

"They were *tourists*, Owen. They go to places so that they can go home and tell everyone where they have been, not that anyone wants to hear about it. It's a particularly valueless kind of status. 'I've kissed the Blarney Stone,' they say. It takes no skill or knowledge to make such a boast, only money. However, Michelle will have her way and will kiss the stone. You can go up the stairs with her, Owen. Don't forget to bring along enough cash for entrance fees and so on."

On our way to Blarney, Michelle tried to jolly Richard into kissing the stone, but he was inflexible.

"You may know the poem about the Blarney Stone," Richard said,

> "There is a stone that whoever kisses,
> Oh! He never misses to become eloquent.
> 'Tis he may clamber to a lady's chamber,
> Or be a member of parliament.

143

Rather ignoble ambitions, both of them. I'll leave the stone to you."

After we came down from the stone, we found Richard in a good mood.

"These ruins are quite nice," he said. "Not all fenced off. It's possible to go anywhere. Contemplating ruins is invariably good for the soul, if only by promoting a salutary humility."

I planned to be on the lookout for signs of humility in Richard.

After a visit to the Blarney Woolen Mills, which Richard endured, we found a place for dinner, where Richard could hold forth.

"Here's a bet, Owen," Richard said, "that your unscrupulous readers can use to take money away from their friends and acquaintances. Take a deck of cards—I happen to have one here—and tell the mark to cut it into four piles. Go ahead, cut. Anywhere you like."

I did.

"Very good," said Richard. "Now ask the mark what he thinks is the chance that at least one of the piles has a spade at the top. If he hesitates, explain to him that the chance that one pile has a spade at the top is $1/4$, since a quarter of the deck consists of spades. Then tell him that because there are four piles and $4(1/4) = 1$, it's almost certain that one of the four piles will have a spade at the top. Of course, you say, it's *possible* that no spade will appear at the top of any of the piles, but it can't be very likely. See what odds he'll take for a series of, say, ten trials. On each trial, he wins if there's at least one spade, and you win if none of the four is a spade. If you settle for 10-1 you'll clean up, because on the average you'll win 3 out of 10 trials."

I turned over the top card on the four piles. "You lose, Richard," I said. "There's a spade."

"Do it nine more times," he said. I did, and Richard won just three times.

"That won't happen in every series of ten trials," he said. "I may be very unlucky and lose all ten. But in this case, as will happen in the long run, I win €10 three times and lose €1 seven times, so I'm €23 ahead. Hand it over."

"No, no, Richard," I said. "Those were hypothetical bets. I'll ask my readers if they can calculate the chance of winning. If you can find those who can't, you can take advantage of them."

"You're a hard man, Owen," Richard said. "Did Socrates dispense tuition for free? Did Plato?"

"Did Socrates make bets?" I asked. "Did Plato?" I had him there.

The professor, beaten for the moment, looked over at a number of books on a bookcase on one side of the room.

"Seeing these books lined up here reminds me of a little curiosity which your readers might enjoy, Owen. There are nine books lined up in a row in this bookcase. If you tried to line them up in all possible arrangements, and if you made one change every minute, working twenty-four hours a day, it would take exactly 252 days to do it. With practice you could no doubt improve to, say, one new arrangement every ten seconds, but it would still take more than a month even without any eating or sleeping. Now increase the number of books to fifteen. If you tried to line them all up in all possible arrangements, making one change per minute, it would take more than 2,486,262 years to do it."

"Let me check that," I said. "It's not that I doubt you, but in mathematics we don't go on authority, we prove all things." All things that we can, I thought, as I took out my calculator. The number of ways of arranging nine objects is $9! = 9 \cdot 8 \cdot 7 \cdot 6 \cdot 5 \cdot 4 \cdot 3 \cdot 2 \cdot 1 = 362,880$. We make a new arrangement every minute. There are 1440 minutes in one day. Therefore in one day we make 1440 arrangements. So the number of days to get all the arrangements is $\frac{362880}{1440} = 252$. For fifteen books we need $\frac{15!}{1440} = \frac{1307674368000}{1440} = 908107200$ days or $\frac{908107200}{365.25} = 2486262.012$ years.

"You're right. You had that number prepared, didn't you?" I said.

"Perhaps," Richard said. "You really shouldn't be so dependent on that calculator, Owen. It stunts the powers of mental arithmetic. You'll notice that I don't carry one."

He had the number prepared. I'm sure of it.

"That reminds me of a little puzzle in permutations that your readers may like," he said. "How many different permutations may be made, each containing ten letters, of the letters in *statistics*, so that no arrangement is repeated?"

"I'll give that to my readers," I said.

Later our conversation turned, for some reason that I cannot recall, to curiosities found in the King James Version of the Bible.

"I've read," I said, "that the shortest verse in the Bible is 'Jesus wept', in John's gospel, and the longest is in the book of Esther."

"That's right," said Richard. "The locations are John 11:35 and Esther 8:9. Your readers may wish to know that the longest name in the Bible is the eighteen-letter name *Maher-Shalaz-Hash-Baz* (Isaiah 8:1). The Old Testament contains 27 books, while the New Testament contains 39, making a total of 66 books. Each of those three numbers is the sum of a square and two cubes: $27 = 5^2 + 1^3 + 1^3$, $39 = 2^2 + 2^3 + 3^3$, and $66 = 8^2 + 1^3 + 1^3$. There are 1189 chapters in the Bible, and 1189 also has that property: it is $26^2 + 1^3 + 8^3$. The number of verses in the Bible is 31102, and $31102 = 63^2 + 14^3 + 29^3$. The two middle verses in the Bible are the 15551st verse from the beginning and the 15551st verse from the end. The number 15551 is not only palindromic, but is $3^2 + 15^3 + 23^3$. The total number of words in the King James Version (according to my files) is 790,871 or, putting it another way, $4^2 + 23^3 + 92^3$. Finally, the number of letters in the Bible is 3,227,553. That is $963^2 + 6^3 + 132^3$. Someone should count the number of syllables to see if that also is a sum of a square and two cubes. Such numbers are not all that common, by the way.

"I mentioned that the number of chapters in the Bible is 1189, which is $118 \cdot 9 + 118 + 9$. By the way, the square of 1189 equals 1,413,721, which is simultaneously square and triangular. It is the sum of the numbers from 1 to 1681. The longest five-vowel word in the Bible with the vowels in any order is *kneadingtroughs* (Exodus 8:3). There are four sixteen-letter words in the Bible: *covenantbreakers* (Romans 1:31), *evilfavouredness* (Deuteronomy 17:1), *lovingkindnesses* (Psalm 25:6), and *unprofitableness* (Hebrews 7:18). Only one word in the Bible contains four *o*s: *footstool* (1 Chronicles, 28:2). The longest chapter in the Bible is Psalm 119, while the shortest is Psalm 117.

"You may have come across the suggestion that Shakespeare secretly had a hand in writing the King James Bible."

"You mean that while Francis Bacon was writing the plays of Shakespeare, Shakespeare was writing the Bible?" I said. "I suppose that he had to occupy his time *somehow*. What's the evidence?"

"It's very compelling to the sort of person who finds such things compelling. The King James Version was completed in early 1611, when Shakespeare was 46 years old. If you look up Psalm 46, you'll find that the 46th word from the beginning is *shake* and the 46th word from the end is *spear*. (The word *selah* at the end is not part of the Psalm.)"

"What's the 46th book of the Bible?" I asked. "Has anyone looked for Shakespeare there?"

"As far as I know he hasn't been found. He must have exhausted his ingenuity by working himself into the Psalms. After all, he was getting old—he died in 1616, only five years after the KJV was completed. Here are two curiosities concerning that year, 1611. Theists (in the Christian tradition) generally believe that there is 1 God, 2 parts to the Bible (Old and New Testaments), 3 Divine Persons in the one God, and 4 Gospels in the Good Book. Place those four digits together to form one number, 1234. Then $1234 + 16^2 + 11^2 = 1611$. What's more, $1234 + 12 \cdot 34 + 1 + 2 - 34 = 1611$. Incidentally, that number 1234 is interesting in its own right. It equals $-1^5 - 2^5 + 3^5 + 4^5$."

"The Bible teaches us that the love of money is the root of all evil," I said. "That may be going a bit far—I can think of some non-monetary evils—but money is certainly interesting."

"It is, indeed, Owen. The reference, by the way, is Timothy 6:10. I'm glad to see that Ireland has adopted the euro, ill-designed as it is."

"How is that?" I asked. "It looks fine to me."

"I meant the units into which it is divided. The U.S. dollar has 50¢, 25¢, 10¢, 5¢, and 1¢ coins, so that 50¢ can be changed in exactly 50 ways, one for each state in the union. (This counts giving change for 50¢ with one 50¢ piece which, if not helpful to the person who wants change, is at least fair.) Euros have 50, 20, 10, 5, 2, and 1 cent coins—what the need is for the 2 cent piece I can't imagine—so there are many more ways of giving change for a 50.

"There are many interesting facts concerning the dollar. The smallest number of coins needed to give exact change for any amount less than a dollar is nine. One half-dollar, one quarter, one dime, two nickels, and four pennies. Alternatively, one could have one half-dollar, one quarter, two dimes, one nickel, and four pennies. It's possible to make change for a dollar with seventy-three coins: three dimes and seventy pennies. Or with seventy-four: two dimes, two nickels, and seventy pennies. Or seventy-five (one 10, four 5s, and seventy 1s) or seventy-six (six 5s and seventy 1s). But it is impossible to make change for a dollar with exactly seventy-seven coins. The largest amount of money one can have in coins and not be able to make change for a dollar is $1.19. This amount is made up of one half-dollar, one quarter, four dimes and four pennies."

"What about the quarter? How many ways can twenty-five cents be changed?"

"A quarter," said the professor, "can be changed in precisely 13 ways, appropriate because the U.S. originally consisted of 13 states."

"What about the dime?" I asked. "Wait, I can get that. A dime is ten pennies, a nickel and five pennies; two nickels, or a dime itself. That makes four. So a dime can be changed in four ways."

"You have it. Four because Independence Day is the Fourth of July."

"How about 75 cents? How many ways can three quarters of a dollar be changed?"

"Exactly 134 ways," Richard said. "13 for the 13 states and 4 for the 4th of July."

"Very nice," I said. What about the dollar itself?"

"A dollar," said Richard, "can be changed in 293 different ways, including the dollar itself."

"Finally," I said, "we are away from American history."

"Au contraire, Owen. $293 = -1 + 7 \cdot 7 \cdot 6$."

"Very nice. But what about the euro? Do you happen to know how many ways a euro can be changed?"

"Certainly," Richard said. "A euro can be changed in precisely 4563 different ways, including the euro itself in the count. It's strange that the U.S. should be involved with that European number, but it is. The U.S. originally consisted of 13 states, there are 32 counties in the island of Ireland, and the U.S. at present consists of 50 states. We have $13 + 32 = 45$, $13 + 50 = 63$, and there, after concatenation, is 4563. Perhaps the U.S. will eventually adopt the euro too.

"The large number of ways a euro can be changed, compared with a dollar, is mostly owing to the fact that there is a coin for .02 of a euro. A friend of mine, Dr. Joe Manning, professor of computer science at University College, Cork, wrote a computer program that gave the number of ways of changing various amounts."

Richard handed me a card.

U.S. Dollar		Euro	
Amount	Number of ways	Amount	Number of ways
5¢	2	2¢	2
10¢	4	5¢	4
25¢	13	10¢	11
50¢	50	20¢	41
$1	293	50¢	451
$2	2728	€1	4563
$5	98411	€2	73682
$10	2103596	€5	6295434
$20	53995291	€10	321335886
		€20	23812353521

"The coins that can be used to make change are 1, 5, 10, 25, 50 cents and $1 for the dollar (though the 50-cent piece and the dollar coin circulate hardly at all) and 1, 2, 5, 10, 20, 50 cents, €1, and €2 for the euro. Since the British pound has coins of exactly the same denominations as the euro, but pence and pounds instead of cents and euros—quite a coincidence, is it not?—the number of ways of changing pounds and pence is exactly the same as the number of ways of changing the corresponding amount in euros. You can see, Owen, from the card that a $20 bill can be changed in over 53 *million* ways, and that a €20 note can be changed in over 23 *billion* ways! Most of your readers will probably find those figures surprising! Incidentally, the figures given for the number of ways of changing various amounts in U.S. currency apply equally to the Canadian dollar."

"That's good. Canadian readers of *The Mathematical Universe* will find this information interesting."

"Your older readers in Ireland may recall the old currency in circulation before decimalization in 1971. There were four farthings (or two half–pennies) in a penny, there was a three-penny piece, a six-penny piece, a shilling (twelve pennies), two shillings, a half crown (30 pennies), a ten-shilling note and the one-pound note. Your readers may be interested to know that the old pound could be changed in 324,946,183 different ways and two pounds could be changed in 30,076,990,438 different ways."

"That's a huge number," I said. "It's a good thing we've got computers."

"Actually, the number can be found fairly easily with pencil and paper. Mathematics, even recreational mathematics, does not really *need* computers, Owen. It got along very well before there were any, and should they all disappear tomorrow, mathematics will carry on. Universities would have to close down their computer science departments, but not their mathematics departments. Mathematics is independent of mechanisms."

"Very true," I said. "Today is the 23rd of the month. Do you have any interesting 23 items? I hope you won't tell me, '23 skidoo'."

"Certainly not, Owen," Richard said. "The number 23 is one of those numbers that seem to turn up in the most unexpected places. For example, human beings have 23 chromosomes. The smallest number of people in a group so that there is a greater than 50 percent chance that at least two people in the group share a common birthday (date and month, ignoring year) is 23. All positive numbers can be expressed as the sum of at most nine positive cubes. Only two numbers actually require nine cubes to represent them. One of these is 239, the other is 23. The number of digits in $23! = 25852016738884976640000$ is 23. Shakespeare's date of birth was April 23, and so was his date of death. In 1932, (note the reversal of 23) a gangster named 'Dutch' Schultz (he was born Arthur Flegenheimer) organized the killing of Vincent 'Mad Dog' McColl in New York City. McColl was murdered on 23rd Street. Schultz was later assassinated in 1935 on October 23rd. His killer, Charlie Workman, was convicted and sentenced to life imprisonment. Workman was paroled after serving 23 years.

"Mathematically, a number of very beautiful properties and equations have been discovered involving the number 23. For instance, $23 = 5 + 7 + 11$. The five digits of that equation from left to right contain the first five primes. The product of the twin primes 3 and 5, plus the sum of those same primes, is 23. A nice expression involving the digits of 23 is $2! + 3! = 2^3$. A factorial expression involving 23 is $1! + (2! + 2!) + (3! + 3! + 3!) = 23$. Also $23 = -(2^2 - 3^3)$. The smallest prime whose reversal equals a power is 23 ($32 = 2^5$). Your readers might enjoy looking for others."

"It's just as well," I said, "that you didn't kiss the Blarney Stone, Richard. If you were any more eloquent I wouldn't be able to keep up with you at all."

"Ah, Owen, you flatter me." said Richard. "I see that you put your trip up to the Stone to good use."

Solutions

1. The deck is cut into four face-down piles and the bet is that none of the cards on the top of the four piles is a spade. Since the mark thinks that this is unlikely to happen because the chance that one pile is topped by a spade is 1/4, there are four piles, and $4(1/4) = 1$, the mark gives odds. We will see what the proper odds are.

   The procedure is effectively choosing four cards from fifty-two. This can be done in

$$\binom{52}{4} = \frac{52 \cdot 51 \cdot 50 \cdot 49}{4 \cdot 3 \cdot 2 \cdot 1} = 270725$$

   ways. The number of ways of choosing four cards none of which is spades is the number of ways of choosing four cards from the 39 non-spades:

$$\binom{39}{4} = \frac{39 \cdot 38 \cdot 37 \cdot 36}{4 \cdot 3 \cdot 2 \cdot 1} = 82251$$

   ways. So, the chance that none of the four cards at the top of the piles is a spade is $82251/270725 = .3038$. That is, about three times out of ten none of the four piles will have a spade at the top. So, if the odds that the mark agrees to are better than 10-3, you have a favorable bet, and the longer they are the better the bet is. If you can talk the mark into 10-1, on the average you will win 23 units over ten trials. Even at 5-1, the expected gain is $10(5(.3038) - 1(.6962)) = 8.22$.

2. How many different permutations of the word *statistics* may be formed, so that no arrangement may be repeated? There is a handy formula that can be used to solve problems of this type. *Statistics* contains the letter *s* three times, the letter *t* three times, the letter *i* twice, and the letters *a* and *c* once each. If we have *n* things, *p* being of one similar category, *q* being of another category, and *r* of a third category, then the total number of ways in which all the *n* objects can be arranged so that no arrangement is repeated is $\frac{n!}{p!q!r!}$. In our problem $n = 10$, $p = 3, q = 3$, and $r = 2$. The formula then gives $\frac{10!}{3!3!2!}$ or 50,400. This is the possible number of permutations

of ten letter arrangements of the word *statistics*, so that no permutation is repeated.

References for further reading

Henry E. Dudeney, *Amusements in Mathematics*. Dover, 1958, puzzle 32: "The Excursion Ticket Puzzle", pages 5 and 151.

Martin Gardner, *The Magic Numbers of Dr. Matrix*. Prometheus Books, 1985, chapter 18, "The King James Bible".

The professor at the university

By pulling strings, I arranged for Richard to be invited to give a talk to an undergraduate mathematics group in Dublin. Actually, I had to pull only one string, suggesting the possibility to someone that I knew. Societies that present talks are always grateful to have speakers, especially those that cost very little. I knew that I had succeeded when Richard called.

"Owen," he said, "My fame has spread at least as far as Dublin. I've been invited to speak there. You'll want to come and listen, won't you? It would be helpful if you could provide transportation. You won't mind if Michelle comes along, will you?"

"Of course not," I said. "What will you talk about?"

"Oh, this and that," he answered. "It's an undergraduate group, so it can't be anything very deep. Undergraduates have short attention spans, so they won't notice if the talk doesn't hang together. And they'll be so happy to have a decent speaker that I could talk about anything."

"You're sure that you'll be more decent than what they usually get?" I asked.

"Owen, you haven't gone to enough mathematics talks. All you have to do to stand out as a terrific speaker is to avoid obvious blunders. Not too long ago I was at a talk where the speaker never looked at the audience because he was looking at the projections of his transparencies. They were in type too small to be seen, so he was reading them to us, word for word, symbol by symbol. Since he wasn't using the microphone, he could hardly be heard, though his shadow could be seen when he stood in the way of the light from the projector. He also went ten minutes past his allotted time. Not only wasn't he hooted from the platform, when he finally finished he got an undeserved patter of applause. I will be splendid in comparison."

The lecture room was about two-thirds full when we arrived, with more empty seats towards the front than in the back. Richard had been prepared for this. Undergraduates, he told us on the way there, have a highly developed fear of fire and always want to be close to the exits, just in case.

Richard started with π, the ratio of the circumference of a circle to its diameter, that crops up almost everywhere in the world of mathematics, outside of circles as well as in them. It starts

$$3.14159265358979323846264338327950288841971\ldots$$

and goes on forever, never repeating and never terminating. It was calculated to more than 206 *billion* decimal places in 1999. That is considerably more than anyone will ever need. Richard mentioned two mnemonic devices for memorizing a few digits of π: *How I wish I could calculate pi* and *May I have a large container of coffee?* (The number of letters in the words in each sentence corresponds to the digits of π.) It is probably easier to remember the seven or eight digits than to recall the mnemonic and count letters, but to recall fifteen digits it may be easier to bring to mind the mnemonic *How I want a drink, alcoholic of course, before reading the heavy chapters in quantum mechanics.* Poetry is easier to remember than prose, and

> Now, I wish I could recollect pi.
> "Eureka," cried the great inventor.
> Christmas pudding, Christmas pie
> Is the problem's very center.

gives twenty-one digits and

> Now, I will a rhyme construct,
> By chosen words the young instruct.
> Cunningly devised endeavour,
> Con it and remember ever,
> Widths in circle here you see,
> Sketched out in strange obscurity.

gives thirty-one, though to be correct we must spell *endeavour* in the British manner.

The professor then told his audience that the millionth digit and the 50 millionth digit of π are both 5 and the 10 millionth and the 100 millionth digits are both 9, where the initial 3 is included. He then gave some fractions that approximate π in which both the numerator and denominator are palindromic.

I had my notepad and pen handy, and managed to note them:

$$\frac{666}{212} = 3.141509\ldots$$

$$\frac{1633361}{519915} = 3.141592\ldots$$

$$\frac{6259909099526}{1992590952991} = 3.141592653589787\ldots$$

He also gave a product, where each number is palindromic,

$$(1.09999901)(1.19999911)(1.39999931)(1.6999961) = 3.141592\ldots$$

As an example of the ubiquity of π, he mentioned the formula for the period of a pendulum. This, he explained, is the time required for one complete vibration; that is, for the pendulum to swing from one crest to the next crest. The formula that he gave is

$$T = 2\pi\sqrt{\frac{L}{g}},$$

where T is the period in seconds, L is the length of the pendulum in meters, and g is the acceleration due to gravity. The value of g in the formula is 9.81 meters per second squared. Thus, the period of a pendulum one meter in length is

$$2\pi\sqrt{\frac{1}{g}} = 2.006066\ldots$$

seconds. Some large clocks, he said, have pendulums of about that length, so the formula can be checked by seeing if they tick once every two seconds.

He went on to give two representations for π:

$$\frac{\pi}{2} = \frac{2\cdot2\cdot4\cdot4\cdot6\cdot6}{1\cdot3\cdot3\cdot5\cdot5\cdot7}\ldots \text{ and } \frac{\pi-3}{4} = \frac{1}{2\cdot3\cdot4} - \frac{1}{4\cdot5\cdot6} + \frac{1}{6\cdot7\cdot8} - \cdots$$

I recognized the first as Wallis's product, but the second was new to me. He mentioned the amazing equation $e^{\pi i} = -1$ and gave two representations for e:

$$e = 1 + \frac{1}{1!} + \frac{1}{2!} + \frac{1}{3!} + \frac{1}{4!} + \ldots \text{ and } e = \lim_{n\to\infty} \frac{n}{\sqrt[n]{n!}}.$$

I again knew the first but not the second.

The professor then moved on to number theory. He recalled the Hardy - Ramanujan story concerning the number 1729, in which Ramanujan instantly recognized that the taxicab number that Hardy had mentioned, 1729, is the smallest integer that can be expressed as the sum of two cubes in two different ways:

$$1729 = 9^3 + 10^3 = 12^3 + 1^3.$$

Richard then stated that the smallest number expressible as the sum of two cubes in three different ways was 87539319:

$$87539319 = 167^3 + 436^3 = 228^3 + 423^3 = 255^3 + 414^3.$$

It was John Leech, said Richard, who discovered this.

Richard then went on to say a few words about number curiosities. The number 5986, he said, was unusual because $5986 = 2 \cdot 41 \cdot 73$, a prime factorization in which each of the nine digits appears just once. Richard then gave another similar equation, but this contained the nine digits and a zero: $28651 = 7 \cdot 4093$. He then casually mentioned that

$$14368485 = 3 \cdot 5 \cdot 17 \cdot 29 \cdot 29 \cdot 67,$$

a prime factorization in which each of the digits appears twice.

The professor said that the universe is younger in seconds than there are permutations of a 52-card deck of cards. He startled many of the audience when he went on to state that the universe is younger in seconds than there are permutations of a 26-card deck. He told his audience that one hundred trillion years after the Big Bang the universe would *still* be younger in seconds than there are arrangements in a deck of 52 cards. I could see that many in the audience were astonished by what they were being told.

The professor then casually mentioned the game of chess and described it as probably the most popular board game in the world. Richard asked if there were any chess players in the audience. At least fifty per cent of the undergraduates raised their hands. I was wondering what was coming next. The professor stunned his audience when he casually mentioned that there were more possible moves in one game of chess than there were atoms in the universe.

He went on to explain that White in chess can make a first move in any one of 20 different ways. Black can then respond in any one of 20 ways. Therefore there are 400 different ways the first two moves in chess can be made. As the game progresses the number of possible moves for both players increases enormously. A typical game consists of approximately 40 moves for each player. The average number of possible moves available at each turn, from the commencement of the game to the end, is about 38. Therefore in the course of the game each player has available about 38^{40} possible moves, so the total number of possible moves for the two players is approximately 38^{80}. This number is in the region of 10^{126}. Scientists estimate that the number of atoms in the observable universe is 10^{79}. Thus, there are far, far more possible moves in a game of chess than there are atoms in the universe.

The professor then said that most people who play bridge are probably unaware of the fact that the total number of bridge hands possible is $\binom{52}{13}$, which equals 635013559600. That's over 635 billion hands! Richard went on to say that if one continuously played bridge, one could reasonably expect to receive a repeat hand (that is, an identical hand dealt at an earlier stage) only after approximately one million hands are dealt. He then told his audience that the total number of *deals* at bridge was much larger. The professor explained that although you could be dealt any one of over 635 billion hands at bridge, the possible number of different hands dealt to the other three players was also enormous.

Richard said that the exact number of deals in bridge is

$$53, 644, 737, 765, 488, 792, 839, 237, 440, 000.$$

This, said Richard, raised an interesting question. What are the odds that all four players at bridge will be dealt a complete suit? The odds of such an event occurring, Richard said, are one chance in

$$2, 235, 197, 406, 895, 366, 368, 301, 560, 000,$$

so that reports of such an event that occasionally appear in newspapers should be taken with several grains of salt. It is almost certain that a hand other than the hand of chance has been operating. Richard went on to say that to appreciate the enormous odds against this happening one should ponder the following result. Assume that the world's population is six billion people. Assume that every member of it dealt a bridge hand every second of every day for one thousand years—no sleeping, no eating allowed. Nothing but rapid shuffling! The odds against such a deal are over 11 million to one.

The professor then returned to number theory and spoke about the square root of 2, whose decimal representation starts with 1.4142135... and goes on forever. It is easy to prove that the square root of 2 cannot be expressed exactly as a fraction. The professor gave such a proof at the lecture. He then projected a transparency with two sequences

$$1, 3, 7, 17, 41, 99, 239, 577 \ldots$$
$$1, 2, 5, 12, 29, 70, 169, 408 \ldots .$$

He explained that the next term in either can be found by doubling the last term and adding the one preceding that, as $17 = 2 \cdot 7 + 3$ and $41 = 2 \cdot 17 + 7$. Place any number, said the professor, in the first sequence over its corresponding number in the second sequence, as $3/2, 7/5, 17/12, 41/29$, and so on. The fractions converge to the square root of 2. For example, $7/5 = 1.4$, $17/12 = 1.4166666\ldots$, $41/29 = 1.4137931\ldots$, $99/70 = 1.4142857\ldots$, and so on. He mentioned another way of generating the fractions. Take any one, divide it by 2 and add its reciprocal. Applying this to $3/2$ gives

$$\frac{3}{4} + \frac{2}{3} = \frac{17}{12} \text{ and then } \frac{17}{24} + \frac{12}{17} = \frac{17 \cdot 17 + 24 \cdot 12}{24 \cdot 17} = \frac{289 + 288}{408} = \frac{577}{408.}$$

Similar things can be done for the square roots of 3, 5, 7, ... he said, but he would leave to the audience the pleasure of finding that out.

The professor next spoke of the standard paper sizes in common use in Europe. Reference numbers such as A3, A4, A5, and so on denote these. He mentioned the curious but little known fact that all these paper sizes have their length and width in the same proportion, and that any particular paper size may be reduced to the next smaller size simply by cutting it in half. For example, to reduce a sheet of A4 paper to the next smaller size, A5, cut the sheet of A4 paper in two halves, cutting from the longest side to the opposite side. If you place two A4 sheets next to each other, you will find that the ratio of the length to the width is unchanged. The principle behind this system, he said, is that the ratio between the length and width of the paper must be in the proportion of $\sqrt{2}$ to 1. With A4 sheets, for example, the length is 297mm and the width is 210mm. The ratio, one of the fractions derived from our sequences, $(297/210 = 99/70)$ is $1.41428\ldots$, quite close to $\sqrt{2}$. In other words, if the width of the paper is 1 unit, then the length of the paper (ideally) equals the square root of two. (Can the reader prove this? The solution is given at the end of the chapter.) An American $8\text{-}1/2''$ by $11''$ sheet of paper, or 216mm by 279mm, is not as long as an A4 sheet, and when cut in half gives two $8\text{-}1/2''$ by $5\text{-}1/2''$ sheets, not a standard size.

Richard then spoke about amicable numbers. These are pairs of numbers where each is the sum of the proper divisors of the other. (That is, the number itself is not counted among the divisors.) The smallest such pair is 220 and 284. The proper divisors of 220 are 1, 2, 4, 5, 10, 11, 20, 22, 44, 55 and 110, which sum to 284. The proper divisors of 284 are 1, 2, 4, 71 and 142, which sum to 220. Other pairs of amicable numbers are 1184 and 1210, and 2620 and 2924. Richard gave some interesting facts about amicable numbers, but I do not have the space here to go into detail.

The professor then spoke of perfect numbers. A perfect number, he reminded them, is a number that equals the sum of all its divisors, excluding itself. The smallest perfect number is 6. The next three are $28 = 1 + 2 + 4 + 7 + 14, 496$ and 8,128. Richard gave some fascinating facts on perfect numbers, but again due to lack of space I cannot give details.

The professor then mentioned multiply-perfect numbers. The smallest such number is 120. Its factors are 1, 2, 3, 4, 5, 6, 8, 10, 12, 15, 20, 24, 30, 40, and 60. These numbers sum to 240, which is twice 120. If 120 is counted as a factor of itself, then the divisors of 120 sum to 360. Hence 120, which is a multiply perfect number, is sometimes called a tri-perfect number. Ordinary perfect numbers could be called bi-perfect numbers, but they aren't.

I knew that this talk of amicable numbers, perfect numbers and multiply-perfect numbers was leading to something, but I did not have a clue as to what it was. At that point I do not think anyone in the audience did either. It was not long before Richard enlightened the audience and me.

The smallest perfect number, Richard reminded us, was 6. The smallest multiply perfect number was 120. The smallest pair of amicable numbers is 220 and 284. Their sum is 504. Is it mere coincidence, asked Richard, that the first perfect number, 6, is $1 \cdot 2 \cdot 3$? Is it coincidental, he asked, that the smallest multiply perfect number, 120, is $4 \cdot 5 \cdot 6$? Is it coincidental, or is there some deep reason, Richard asked, that 504, the sum of the numbers in the first amicable pair, is $7 \cdot 8 \cdot 9$? What, he asked, could be lurking behind $10 \cdot 11 \cdot 12 = 1320$ besides the fact that its digits are a permutation of 0, 1, 2, 3? There must be something.

The professor then spoke about his interest in wordplay. The word *testament*, he said, is defined as a written statement of one's belief. Is it not curious, he asked, that *testament* is an anagram of *statement*? Is it not curious, he asked, that the name *William Shakespeare*—one of the greatest literary geniuses of all time—is an anagram of *I am a weakfish speller*?

Most people, said the professor, usually find coincidences interesting. Some wonder whether some coincidences are really random occurrences, or if they are part of a deeper, unrevealed plan of nature. Is it coincidental, he asked, that water in its frozen state floats in its own liquid? If water did not have this property life could not exist on this planet. Water, the professor said, is virtually unique among liquids in possessing this property.

Apparent coincidences, Richard said, can lead humans on to great discoveries. The apparent coincidence that mammals have approximately the same ratio of salt in their blood as there is salt in seawater strengthened the belief that life originated in the sea. The apparent coincidence that the map of South

America fits snugly into the map of Africa reinforces the belief in the theory of plate tectonics.

There are numbers, the professor said, that somehow or another repeatedly crop up in nature. The number 10^{40} is one such. For example, the size of the observable universe divided by the radius of an electron is approximately 10^{40}. The ratio of the electromagnetic and gravitational forces between two protons is approximately 1 to 10^{40}. The total number of protons in the observable universe is approximately 10^{80}, the square of 10^{40}. Are the appearances of these numbers merely coincidental, asked the professor, or is nature through these numbers trying to tell us something?

The five most important quantities in mathematics, the professor said, are the square root of -1, 0, 1, e and π. All of them are indispensable in mathematics. The square root of -1 is an imaginary number. Is it coincidental, asked the professor, that the other four numbers are all less than four?

Finally, Richard referred to the famous formula $e^{\pi i} + 1 = 0$ and described it as perhaps the most beautiful equation in the whole of mathematics. He said that when he had encountered that equation for the first time many years previously, knowing what the various symbols represented, he had felt that a deep secret of nature had been revealed to him. He said that he still experienced feelings of humility, awe, and gratitude when he contemplated the equation.

Richard reminded his young audience that they were starting out on the winding road of life, which would have many pitfalls. As you go through life, he told them, try to help others along the way. If you want to be happy, he said, remember what James Barrie said: "Those who bring sunshine to the lives of others cannot keep it from themselves." Remember it, he said, and try to live by it. If you do, he said, you would not go too far wrong.

Richard ended the lecture by reminding the audience that the first step to knowledge is to realize how ignorant we are. What we know about anything, he said, is only a tiny fraction of what we do not know.

The professor thanked the undergraduates for their time and attention, and encouraged them to continue their enquiries in to the deep, beautiful, and astonishing mathematical structure that exists outside of human minds.

The professor received an enthusiastic round of applause from the undergraduates, and indeed from many of the professors in the lecture theatre.

Later we went to visit my brother, Dominic, at his restaurant, O'Shea's Traditional Irish Restaurant. If any of my readers ever happen to be in Dublin, it can be found at 23 Anglesea St., Temple Bar. Irish stew, of course, is on the menu, as is a Guinness beef casserole. I usually go for the steak and chips or the mixed grill. As we were unwinding I said, "That was a great lecture you gave, Richard. The audience were enthralled."

"Thanks," said Richard. "They were a good audience. It can be deadly when the audience isn't coming along with you, at least half-way. You noticed that I ended three minutes early? That will guarantee that everyone will think it was a good talk."

"You grabbed their attention," said Michelle, "when you mentioned that the universe was younger in seconds than there are arrangements in a deck of cards. Just when they were trying to take in that astonishing fact, you tell them that the universe is younger in seconds than there are arrangements in half a deck. Did you think of asking them how large a deck would be needed for its number of arrangements to equal the age of the universe in seconds?"

"That might have distracted them," Richard said. "Those facts, incidentally, are easily proved. The number of arrangements in a deck of 52 cards is 52!, about $8.0658 \cdot 10^{67}$. The universe is at most sixteen thousand million years old, or $16 \cdot 10^9$ years. A day consists of 86,400 seconds. There are 365.25 days approximately in one year. Call a year 366 days. Thus there are approximately $366 \cdot 86400$, or 31622400 seconds in one year. In sixteen thousand million years there are approximately $505958400 \cdot 10^9$ or approximately $5.1 \cdot 10^{17}$ seconds. That is the approximate number of seconds that have elapsed since the beginning of the universe. The number of ways that a 52-card deck can be arranged is $8.0658 \cdot 10^{67}$. This last number is greater than $5.1 \cdot 10^{17}$ by a factor of $1.59 \cdot 10^{50}$. So the number of arrangements in a 52-card deck of cards far exceeds the age of the universe in seconds. The number of arrangements in a 26-card deck of cards is 26!, or roughly $4.03 \cdot 10^{26}$, which is more than $790,000,000$ times greater than $5.1 \cdot 10^{17}$. In other words, the number of ways a 26-card deck of cards can be arranged is over 790 million times greater than the age of the universe in seconds. One hundred trillion years—that's 10^{14} years—after the Big Bang the number of seconds that will have elapsed since the beginning of the universe is less than $3.2 \cdot 10^{21}$. This is far, far less than $8.06 \cdot 10^{67}$, which is the number of different ways a 52-card deck can be arranged. Even a 22-card deck has more arrangements than the number of seconds in the age of the universe. The universe is really quite young."

"*Ars longa, universitatitum brevis*, as it were?" I put in, pleased with my cleverness.

"Owen, you are a Celt," he said. "Stay away from Latin. That phrase that came so trippingly and unfortunately from your tongue has 29 letters, which reminds me that the human skull has 29 bones and that approximately 29 per cent of the earth's surface area is land. All the months of the Jewish and Muslim calendars have either 29 or 30 days. A leap year February in the Gregorian calendar has 29 days. The highest possible score for a hand in a game of cribbage is 29. With just seven cuts $(-2 + 9)$ it is possible to divide a cake into 29 pieces. The metallic element copper has atomic number 29. The atomic bomb dropped on Hiroshima on August 6, 1945, was released from an airplane known as a B-29. The horrendous bombing in Omagh, Northern Ireland, on August 15, 1998, killed 29 people. (Two unborn babies also died in the attack.)"

"Yes, Richard, I recall that bombing. Another appalling example of man's inhumanity to man. What about the number 29 itself? Any interesting information on that?"

"It's prime, of course," Richard said, "and it's the smallest prime of the form $p^p + 2$, where p is prime. Also, $29 \equiv 2 \pmod 9$."

"Just a second," I said, still feeling bright. "$23 \equiv 2 \pmod 3$."

"True," said Richard, "but congruences modulo 9 are three times more unlikely than congruences modulo 3."

I let him have his way and didn't mention that $49 \equiv 4 \pmod 9$. It might have distracted him.

"The seventh Lucas number is 29. The first four primes can be arranged in five ways to equal 29: $2 \cdot 7 + 3 \cdot 5, 5 \cdot 7 - 2 \cdot 3, 27 + 5 - 3, 25 + 7 - 3$, and $72/3 + 5$. The sum of 2^2, 3^2 and 4^2 is 29, and $2 + 9 + 2 \cdot 9 = 29$. Also, 29 is one less than 30, which is what journalists put at the end of their stories. That is what I am going to use 29 + 1 for now: Michelle and I will be leaving Ireland the day after tomorrow. Now don't carry on, Owen—we will be back, though we don't know exactly when. You'll just have to keep your column going without me."

"I'll try," I said.

Solutions

1. How are the odds for the perfect deal at bridge calculated? You can be dealt any one of $\binom{52}{13}$ hands at bridge. Each of these hands can be combined with any one of the $\binom{39}{13}$ hands that may be dealt to, say, your left hand opponent. Your partner may now get any one of $\binom{26}{13}$ possible hands. The fourth player must take the thirteen cards left. We can express these results as

follows:

$$\text{Possible hands for Player 1}: \frac{52!}{13! \ 39!} = 635,013,559,600$$

$$\text{Possible hands for Player 2}: \frac{39!}{26! \ 13!} = 8,122,425,444$$

$$\text{Possible hands for Player 3}: \frac{26!}{13! \ 13!} = 10,400,600$$

$$\text{Possible hands for Player 4}: 1$$

When these four numbers are multiplied, one obtains the total number of possible deals,

$$\frac{52!}{13! \ 13! \ 13! \ 13!} = 53,644,737,765,488,792,839,237,440,000.$$

There are 4! or 24 ways in which each player can be dealt a complete suit. Divide the 29-digit number above by 24, and one obtains 2,235,197,406,895,366,368,301,560,000. The probability therefore that all four players at bridge will be dealt a complete suit is one chance in 2,235,197,406,895,366,368,301,560,000.

To work out the odds quoted by the professor, do the following. Divide the number above by 6,000 million. Divide the result by 86400, which is the number of seconds in a day. Divide that result by 365.25, which the average number of days in a year. Then divide that result by 1,000. The answer is a little more than 11,804,855.28, as in the odds that the professor gave.

2. We were told that if one wishes to reduce a sheet of A4 paper to the next smaller size, (which is A5) simply cut the sheet of A4 paper in two halves, cutting from the longest side to the opposite side. The principle behind this system, we were told, is that the ratio between the length and width of the paper must be in the proportion of $\sqrt{2}$ to 1. (With A4 sheets, for example, the length is 297mm and the width is 210mm.)

Here is the proof of this. Let l be the length of the sheet of A4 paper, and w its width. The length of A5 paper then is w and its width is $0.5l$. The proportion of length to width is equal in both sheets. Therefore we can write

$$\frac{l}{w} = \frac{w}{0.5l}, \quad \text{or} \quad 0.5l^2 = w^2, \quad \text{or} \quad l^2 = 2w^2, \quad \text{or} \quad \frac{l^2}{w^2} = 2.$$

So, $\dfrac{l}{w} = \sqrt{2}$.

References for further reading

Martin Gardner, *Mathematical Magic Show*. Penguin Books, 1985, Chapter 12, "Perfect, Amicable, Sociable".

John Haigh, *Taking Chances*. Oxford University Press, 1999, Chapter 13, "Lucky for Some—Miscellania".

About the Authors

Owen O'Shea was born in 1956 in Cobh, County Cork, Ireland. Owen (and his twin brother, Michael) was the youngest of a family of eleven children. Owen has said that his parents and family lived on a modest income, but that his parents were hardworking and honest. His father was a non-commissioned officer in the Irish Naval Service. Owen's mother was a psychiatric nurse in London before she was married.

Owen O'Shea is a single person. He is employed as a civilian employee in the Department of Defense in Ireland. In his younger days Owen lectured occasionally on recreational mathematics to university students. He also frequently at that time voluntarily assisted various people in his community (particularly senior citizens) to complete forms in the areas of social welfare, pensions and tax to ensure that these persons obtained their legitimate entitlements from the Irish State.

Owen is the author of a number of newspaper articles that appeared in Ireland in recent years. These articles ranged from such diverse topics as to how the date of Easter is calculated to a biographical sketch of Martin Luther King Jr. Owen has a wide range of hobbies. His first love is, of course, recreational mathematics. He is also interested in collecting and spotting strange coincidences. His other interests include mathematical magic, science, astronomy, nature, philosophy, poetry, conjuring, word play and history.

This is Owen's first book. He says that he will be very pleased if those who read this book enjoy it as much as he has enjoyed writing it.

Underwood Dudley earned his B.S. and M.S. degrees from the Carnegie Institute of Technology and his doctorate (in number theory) from the University of Michigan. He taught briefly at the Ohio State University and then at DePauw University from 1967–2004. Woody has written six books and many papers, reviews, and commentaries. He has served in many editing positions, including editor of *The Pi Mu Epsilon Journal*, 1993–96 and *The College Mathematics Journal*, 1999–2003. He is widely known and admired for his speaking ability—especially his ability to find humor in mathematics. He was the PME J. Sutherland Frame lecturer in 1992 and the MAA Pólya lecturer in 1995–96. Woody's contributions to mathematics have earned him many awards, including the Trevor Evans award, from the MAA in 1996, the Distinguished Service Award, from the Indiana Section of the MAA in 2000, and the Meritorious Service Award, from the MAA in 2004.

Index

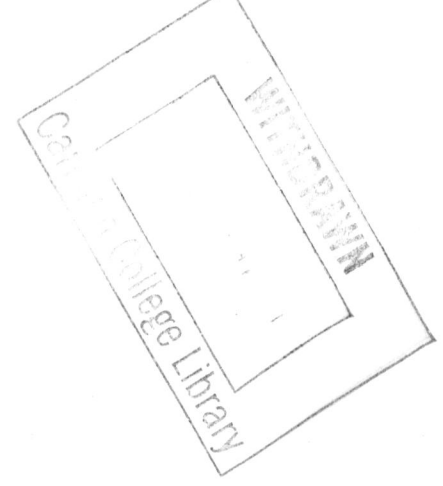